Diversity Of Dragonflies And Damselflies Insecta Odonata

By

A. Joshua Andrews

CONTENTS

LIST OF TABLES

20	Dragonfly population density for Sholavaram in the year 2016
21	Students t test analysis for dragonfly population density at Sholavaram
22	Different superscripts for dragonfly population density in the same column showing significant difference at P< 0.05 level at Avadi in the year 2015
23	Dragonfly population density for Avadi in the year 2015
24	Different superscripts for dragonfly population density in the same column showing significant difference at P < 0.05 level at Avadi in the year 2016
25	Dragonfly population density at Avadi in the year 2016
26	Students t test analysis for dragonfly population density at Avadi
27	Different superscripts for dragonfly population density in the same column showing significant difference at P< 0.05 level at Ambattur in the year 2015
28	Dragonfly population density at Ambattur in the year 2015
29	Different superscripts for dragonfly population density in the same column showing significant difference at P< 0.05 level at Ambattur in the year 2016
30	Dragonfly population density at Ambattur in the year 2016
31	Students t test analysis for dragonfly population density at Ambattur
32	Different superscripts for dragonfly population density in the same column showing significant difference at P< 0.05 level at Korattur in the year 2015
33	Dragonfly population density at Korattur in the year 2015
34	Different superscripts for dragonfly population density in the same column showing significant difference at P< 0.05 level at Korattur in the year 2016
35	Dragonfly population density at Korattur in the year 2016
36	Students t test analysis for dragonfly population density at Korattur
37	Different superscripts for dragonfly population density in the same column showing significant difference at P< 0.05 level at Poonamallee in the year 2015
38	Dragonfly population density at Poonamallee in the year 2015
39	Different superscripts for dragonfly population density in the same column showing significant difference at P< 0.05 level at Poonamalee in the year 2016
40	Dragonfly population density at Poonamallee in the year 2016
41	Student t test analysis for dragonfly population density at Poonamallee

42	Different superscripts for damselfly population density in the same column showing significant difference at P< 0.05 level at Poondi in the year 2015
43	Damselfly population density at Poondi in the year 2015
44	Different superscripts for damselfly population density in the same column showing significant difference at P< 0.05 level at Poondi in the year 2016
45	Damselfly population density at Poondi in the year 2016
46	Students t test analysis for damselfly population density at Poondi
47	Different superscripts for damselfly population density in the same column showing significant difference at P< 0.05 level at Puzhal in the year 2015
48	Damselfly population density at Puzhal in the year 2015
49	Different superscripts for damselfly population density in the same column showing significant difference at P< 0.05 level at Puzhal in the year 2016
50	Damselfly population density at Puzhal in the year 2016
51	Students t test analysis for damselfly population density at Puzhal
52	Different superscripts for damselfly population density in the same column showing significant difference at P< 0.05 level at Sholavaram in the year 2015
53	Damselfly population density at Sholavaram in the year 2015
54	Different superscripts for damselfly population density in the same column showing significant difference at P< 0.05 level at Sholavaram in the year 2016
55	Damselfly population density at Sholavaram in the year 2016
56	Students t test analysis for damselfly population density at Sholavaram
57	Different superscripts for damselfly population density in the same column showing significant difference at P< 0.05 level at Avadi in the year 2015
58	Damselfly population density at Avadi in the year 2015
59	Different superscripts for damselfly population density in the same column showing significant difference at P< 0.05 level at Avadi in the year 2016
60	Damselfly population density at Avadi in the year 2016
61	Students t test analysis for damselfly population density at Avadi
62	Different superscripts for damselfly population density in the same column showing significant difference at P< 0.05 level at Ambattur in the year 2015

63	Damselfly population density at Ambattur in the year 2015
64	Different superscripts for damselfly population density in the same column showing significant difference at P< 0.05 level at Ambattur in the year 2016
65	Damselfly population density at Ambattur in the year 2016
66	Student t test analysis for damselfly population density at Ambattur
67	Different superscripts for damselfly population density in the same column showing significant difference at P< 0.05 level at Korattur in the year 2015
68	Damselfly population density at Korattur in the year 2015
69	Different superscripts for damselfly population density in the same column showing significant difference at P< 0.05 level at Korattur in the year 2016
70	Damselfly population density at Korattur in the year 2016
71	Students t test analysis for damselfly population density at Korattur
72	Different superscripts for damselfly population density in the same column showing significant difference at P< 0.05 level for Poonamallee in the year 2015
73	Damselfly population density at Poonamallee in the year 2015
74	Different superscripts for population density in the same column showing significant difference at P< 0.05 level at Poonamallee in the year 2016
75	Damselfly population density at Poonamallee in the year 2016
76	Students t test analysis for damselfly population density at Poonamallee
77	Regression analysis for the dragonfly in the year 2015
78	Regression analysis for the dragonfly in the year 2016
79	Regression analysis for the damselfly in the year 2015
80	Regression analysis for the damselfly in the year 2016

1.0. INTRODUCTION

Biodiversity encloses the mixture of life in all levels of system, classified both by evolutionary and ecological standard. All species have specific role in the ecosystem. The enormous diversity of life on the universe has contributed to human need for many years. The diversity of all species together forms organization which supports the improvement of biodiversity. Through various studies, science has developed techniques to categorize the variability in earth for many years which helps us to estimate the association in the environment. Study on diversity of particular species which comprise of species richness, diversity of species and distribution of individuals among the ecosystem helps us to know all the important aspect of that species. All species together form a healthy ecosystem and they are like a thread holding together. If a species disappears from specific habitat then the unified ecosystem will be affected. Biodiversity is important to obtain healthy environment and sustainable development. The diversity can be attributed to the vast variety of land forms and climates resulting in habitats ranging from tropical to temperate and from alpine to desert. But in the recent days, most of the species are under danger due to industrialization and different kinds of anthropogenic activities which will certainty have an adverse effect on the environment. Hence, the importance to study the diversity of species and to know its functional aspect on the environment is the need of the hour.

Odonates (Odonata: Insecta) have inhabited the earth for around 245 million years and are believed to be the descendants of the Protodonata that existed about 300 million years back. There are 6,256 species of Odonates under 686 genera and 39 families recorded from the world (Schorr and Paulson, 2017). Of this, 487 species and 27 sub species from 152 genera and 18 families are known from India (Subramanian and Babu, 2017). The order Odonata comprises of dragonflies and damselflies divided into three suborders, *viz.*, Anisoptera, Zygoptera and Anisozygoptera. Anisoptera houses dragonflies which are the robust built ones, besides being strong fliers. Eyes meet mid dorsally and occupy the largest part of head. Labium with middle lobe is deeply cleft. The hind wings are broader basally and are held horizontally. Male possesses two superior and one inferior anal appendages and penis is jointed. Ovipositor is normal or atrophied. Oviposition is exophytic. Zygoptera includes damselflies which are slender and weak fliers. Eyes are button like which are much smaller and do not meet. Labium will be variable. Wings are identical and are held vertically at rest. Male possesses two superior and two inferior anal appendages and penis is not distinctly jointed. Ovipositor complete and oviposition is endophytic. Anisozygoptera contains combination of both and are represented by only a few species under a single genus all over the world (Mitra and Mitra, 2009).

The body of an odonate is divided into three parts *viz.*, head, thorax and abdomen. The head is broad, and can be rotated sideways, downwards, forwards and backwards. The head consists of biting mouth parts suitable for predatory life and has a well-developed pair of compound eyes which provides great vision, nearly (360°). In the Zygoptera, these eyes are extensively separated, but in the Anisoptera they meet in the middle line. The ocelli, or simple eyes are seen on the vertex, the area between the eyes. The ocelli help to see the objects which are close. It consists of small antennae which has three to seven segments. The mandibles are strong with powerful teeth and the maxillae bear a lobe like unsegmented palp; the labium modified as lateral lobes and hooks forms the labial palps.

The middle part of the body is known as thorax and comprises of three segments. The first segment or pro thorax consists of anterior, middle and posterior lobes. Prothorax are used for rotating head. The meso and meta thorax merge to form synthorax. The legs are attached anteriorly below the thorax which helps to hold the prey close to its mouth. The wings are attached to the mesothorax and metathorax with markings on the thorax which are used to identify the species. Two pairs of wings are attached to the last two thoracic segments. The wings are made up of numerous veins enclosing cells. The arrangement of the veins and cells are of vital help in identification of odonates. The fore wings are identical and hind wings are held together vertically over the body at rest in damselflies. In dragonflies, the fore wings are dissimilar to the hind wings and they spread the wings at right angle to the body when they are at rest. The wings will be either transparent or colored and venations are extremely specialized.

In most of the anisopterans, the abdomen is either long and cylindrical or dorsoventrally flattened made up of ten segments. The ventral side of the segments two and three will form secondary (accessory) genitalia in male. In few anisopterans, in addition to it, a ventrolateral process known as auricle or oreillets are also seen. Behind the tenth segment, a pair of unsegmented superior anal appendages are present and they are the modified form of cerci. A pair of inferior anal appendages is seen in male zygopterans as while a single median inferior appendage is present in male anisopterans. In male, the gonopore opens ventrally on the ninth segment and while bending the abdomen sperm is transferred from this place to the storage vesicle on the secondary genitalia present in the third segment. Penis gets extended from the second segment and a pair of hamules is seen on either side of penis which helps to hold and guide the female genitalia during copulation. In female zygopterans and in a few anisoptera, a developed ovipositor and a terebra are seen, which are used for cutting, piercing and sawing the tissues of submerged aquatic plants which will help to insert the eggs into the plant tissues which is known as endophytic oviposition. In female anisopterans there is a small ovipositor called vulvar scale but in few species the ninth and tenth segment of sterna forms a cavity to temporarily store the egg masses released from the gonopore.

The odonata life history in closely associated with water bodies. They are adapted to large range of flowing and stagnant water sources. The odonates are found near water bodies like lakes ponds, streams, rivers, tanks, marshes, temporary ponds, pools as well as brackish water. Most of the species are habitat specific and they play a vital role in the aquatic ecosystem. Mating usually occurs in the breeding site which was their primary habitat. The male curves its abdomen downward to deposit the sperms from the genital pore to the accessory genitalia which is on the ventral side of abdominal segments two and three. At the same time the male zygopteran will catch the anterior part of the female prothorax, while the male anisopteran holds the female head. Later they will fly together in a 'tandem position'. In the next stage, a wheel shape is formed in which female bends the abdomen down and makes in such way that its genitalia is in connection with male genitalia. Mating occurs for a few seconds to several hours. Oviposition is followed by copulation. During oviposting, in maximum cases the male continues with the female in tandem position. Nearly all of the odonates lay their eggs in large range of aquatic environment. Female odonates lay eggs from a few hundred to thousands in number where they lay in bundles, in rows or in an irregular pattern. Odonates are hemimetabolous and hence they have a three staged life cycle, *viz.*, egg, nymph and adult. Hatching out of nymph from egg differs from few days to months based on the species and temperature. The first hatched nymph lives on the yolk maintained in the body. The instar period may differ from nine to sixteen based on the species, temperature and before the final instar development of imago occurs. Larva is a voracious feeder and feeds on tadpoles, small fishes and larva of aquatic insects. The development of larva takes weeks and days prior to emergence, the larva finishes its feeding thus the last moult takes place within the larval skin. Emergence occurs on plants which are above the water or near the shore, thus they leave the water edges as matured adults. The odonates rise up by splitting the dorsal side of head and thorax of larva. The newly emerged or teneral dragonfly can be identified by appearance of soap bubble glittering and dim color. Males establish territories while female come to territories only for mate and oviposit. In the life history of insects, Odonata biomass regularly exceeds the other taxa which cross the aquatic terrestrial boundary (Bried, 2005), since they have an ecological significance as a predator and prey in the ecosystem. Nymph and adults exhibit predatory mode of life. Odonata have large eyes which possesses numerous ommatidia specializes for the movement and detection. Most of the little insects are caught by using mouth parts and in certain cases legs are used for capturing the prey. Odonata feed on mosquito, butterflies, moths, midges where the nymph food includes blood worms, water fleas, tadpoles and larvae of mosquitoes or water insects. They are the best natural control agent over some pests and disease causing insects. Odonates feed on living prey all over their adult life. They are also introduced in biological control programs as they feed on many harmful insects of crops and forests. They are very good indicators of freshwater pollution and they can serve as important bioindicators for water quality (Vorster *et al.*, 2020). Odonates play an important role in maintaining the natural balance of ecosystem and used as an indicator of the environmental changes. In scientific research biodiversity, indicators are used as quantifiable environmental factors. Cameron (2002) and Hooper *et al.*, (2005) affirmed that the relationship between biodiversity and the ecosystem functions is a topic that has generated intensive research and controversy in the recent years. Habitat deterioration and pollution impose great pressure on the species and their habitat and odonates has been used as indicator to understand the status of biodiversity in freshwater ecosystem (Suhling *et al.*, 2006; Clausnitzer, 2003). The resulting disruption of ecological processes affects intra and interspecific relationships, ecological services, and many other aspects of ecosystem structure and function (Chesson, 2018: Pereira *et al.*, 2019). Odonata (dragonflies and damselflies) are taxonomically well known and globally recognised for their environmental indicator potential (Bouwman, 2018). The various issues on biodiversity and conservation have been sourced for the current study of odonates.

2.0. REVIEW OF LITERATURE

2.1. Diversity studies of odonates

Fraser (1931, 1933, 1934 and 1936) published three volumes on Odonata in the Fauna of British India including 536 species and subspecies of Odonata, which included 96 species of (Zygoptera and Anisoptera) from Tamil Nadu. Prasad and Singh (1994) studied fauna of conservation areas series, Rajaji National Park. In the study, 38 species from eight families were recorded in Rajaji National Park. Mitra (1995) studied Indravati tiger reserve forest and reported twenty two species of odonates of which ten species were new records. Prasad and Varshney (1995) made a checklist of Indian odonates, and reported 499 species of odonates belonging to three suborders, 17 families and 139 genera.

Prasad (1996) reported 83 species of odonates from Maharashtra state. Samway and Steytler (1996) recorded 26 species of odonates including 13 zygopterans and 13 anisopterans from Dorp spruit river, South Africa. Tyagi (1997) reported a checklist of 491 species of odonates from India. Study focused on the geographical distribution and habitat specific of a particular region (Emiliyamma and Radhakrishnan 2003, Emiliyamma *et al.*, 2005; Subramanian 2005, 2007). Andrew and Tembhare (1997) studied odonates from Vidharbha region and reported 43 species.

Gunathilagaraj *et al.*, (1999) reported 16 species of odonates from rice fields of Coimbatore of which nine were dragonflies and seven are damselflies. Of the 16 species, *Orthetrum Sabina* (dragonfly) and *Ishnura delicate* (damselfly) were the most abundant species recorded in the rice fields. Uniyal *et al.*, (2000) reported 14 species of odonates from great Himalayan national park, Himachal Pradesh. Norma-Rashid *et al.*, (2001) studied freshwater swap lake of Tasek Bera, Malaysia in which fifty nine species of odonates were recorded. Prasad and Kulkarni (2001) had reported 71 species of odonates from Nilgiri Biosphere Reserve, Western Ghats.

Prasad and Kulkarni (2002) stated out of 5000 species from the world, 500 species belonging to 139 genera of 17 families has been recorded from India. Mitra (2002) recorded 205 endemic species of the Odonata from India. Talmale and Kulkarni (2003) recorded

5

19 species of Odonata from paddy field of Bhandara district, Maharashtra. Garcia and Klaas-Douwe (2004) studied Odonata from Ankarafantsika national park. The study recorded 164 individual belonging to 33 species. The study area is huge part of dry forest in Madagascar's Region. Endemic species observed are found common in the Island. Kulkarni *et al.*, (2004) recorded 38 species of odonates from Pench national park, Nagpur, Maharashtra.

Kumar (2005) studied odonates from Khajjiar lake area in Himachal Pradesh and reported 88 species under 52 genera belonging to Zygoptera (31 species) and Anisoptera (57 species). Kandibane *et al.*, (2005) reported from irrigated rice fields of Madurai in Tamil Nadu, nine species of anisopterans and three species of zygopterans. In the study, *Pantala flavescens* and *Diplocodes trivialis* from Anisoptera showed dominance while *Agriocnemis femina* from Zygoptera showed dominance. Emiliyamma (2005) from Kottayam district of Kerala recorded 31 species of odonates belonging to 22 genera and seven families. Kulkarni and Prasad (2005) studied odonates fauna of Melghat Tiger Reserve, Amravati, which comprises of 400 specimens from 24 species of odonates recorded from Melghat Tigar Reserve. *Trithemis aurora* showed dominance with 129 specimens. Subramanian and Sivaramakrishanan (2005) studied the diversity and distribution of stream insect locality in three various habitats and 33 micro habitats by the data obtained from 39 territory of the Western Ghats. The habitats for aquatic insect communities of river ecosystems can be seen within various scales. The habitat and micro habitat distribution helps in understanding the distribution of stream insects and will help to develop bio-monitoring tools over it. Structurally complex micro habitat harbors more damselflies. The diversity and abundance of odonates may vary among various habitats.

Kulkarni *et al.*, (2006) studied fauna of Tadoba Andhari Tiger Reserve, Chandrapur, comprising of 41 species of odonates observed for first time. Mitra (2006) studied the current status of the odonates of Bhutan which had a checklist of four new records with 61 individual of odonates from 16 species in 13 genera and four families being recorded. Wahizatul-Afzan (2006) studied diversity and distribution of odonates in Sekayu recreational forest, Terengganu wherein 593 specimens from 44 species belonging to 11 families were recorded. Zygopterans were recorded with more abundance than the anisopterans, but the family Libellulidae showed dominance.

Papazian *et al.*, (2007) studied the odonates from the French Pacific island territories of Wallis and Futuna, wherein eight anisopterans and two zygopterans species were recorded. Sethy and Siddiqi (2007) observed odonates in Similipal Biosphere Reserve, Mayurbhanj, North Orissa which revealed a distribution of 16 species of Odonates under 14 genera and six families. Of these six families Libelluiidae (six species) was well represented followed by Coenagrionidae (five species), Lestidae (two species), Protoneuridae (one species), Aeshnidae (one species) and Calopterygidae (one species). Sato and Riddiford (2007) studied odonates of S'Albufera natural park, Mallorca wherein 10 species of dragonflies and four species of damselflies were recorded. Sharma and Joshi (2007) studied the diversity of odonates (Insecta) from Dholbaha Dam, Punjab Shivalik, India which recorded 30 species from seven families with the order of dominace with family Libellulidae followed by Coenagrionidae, Aeshnidae, Calopterygidae, Chlorocyphidae, Euphaeidae and Gomphidae. Sharma *et al.*, (2007) studied diversity of dragonflies and damselflies in certain provenance of sandal in India. In the study 672 individuals from 21 species under four families observed in the six provenances of sandal in the study which showed 20 species from Marayoor, 17 species from Bengaluru, 16 species from Javadis, 12 species from Thangali, 11 species from Mandagadde and 8 species from Chitteri. The family Libellulidae showed dominance in all the study area.

Subramanian *et al.*, (2008) studied odonates as indicator of riparian ecosystem health as a case study from south western Karnataka, The significance of riparian land in the odonates diversity was studied wherein 55 species from 12 families were observed. Ebrahimi *et al.*, (2009) studied dragonflies (Odonata) from South-Eastern Iran and reported 27 species of odonates belonging to eight families in Kerman province. The family Libelullidae has dominated the majority of dragonfly's genera and species (55%) in the region and *Trithemis festiva* is a common species in the most of the collecting sites

Fulan *et al.*, (2010) studied the abundance and diversity of odonates for four years after construction of a reservoir in Guadiana River, Spain wherein the study recorded 11 zygopterans and 10 anisopterans representing six families. Singh *et al.*, (2010) recorded 23 species of odonates in which 15 species from seven genera and three families of dragonflies were recorded from Kane wildlife Sanctuary, Arunachal Pradesh, Northeastern India. Rangnekar *et al.*, (2010) reported 66 species of odonates from Goa of which 34 species were newly recorded. In Zygoptera, the family Coenagrionidae dominated the list with 14 species, while in Anisoptera, the family Libellulidae was found in large with 32 species. Of all the families, the maximum contribution to the species diversity was found in the family Libellulidae which accounts to 47.76% followed by Coenagrionidae which contributed 20.89% of the total species. Tiple *et al.*, (2010) studied diversity of odonates in Kanha national park, Madhya Pradesh, India which recorded a total of 36 species of odonates belonging to two suborders and seven families, *viz.*, Gomphidae (two species), Aeshnidae (two species), Libellulidae (twenty species), Coenagrionidae (eight species), Protoneuridae (two species), Euphaeidae (one species) and Lestidae (one species) were recorded. Odonates are good indicators of environment as they are sensitive and directly affected by micro levels changes in habitats in relations to the weather conditions. Arulprakash and Gunathilagaraj (2010) studied abundance and diversity of odonates in Salem districts of Tamil Nadu wherein 21 species of odonates from 13 temporary water bodies in which 14 species from Anisoptera and seven species from Zygoptera which belong to 17 genera and four families from Coimbatore and Salem districts in Tamil Nadu were recorded.

Dow *et al.*, (2012) collected 218 individuals of odonates which included 50 species from nine families in Sungai Bebar, Pahang, Malaysia, which recorded 28 and 22 species of dragonfly and damselfly respectively. Tiple (2012) studied odonates diversity in Achanakmar- Amarkantak Biosphere Reserve, Madhya Pradesh and Chhattisgarh, India, where 70 species of odonates belonging to 12 families were observed. The maximum number of odonates was observed from the family Libellulidae with 31 species, 15 species from family Coenagrionidae, five species from family Gomphidae, three species from Protoneuridae and Lestidae, four species from Aeshnidae, two species from Platycnemididae, Calopterygidae and Chlorocyphidae, one species each from Euphaeidae, Corduliidae and Macromiidae. Out of 70 species, 24 were very common, 23 common, 18 rare, and 5 very rare. Tiple (2012) reported a total 64 species of odonates belonging to 41 genera and nine families from Tadoba national park, Chandrapur district, Maharashtra, India. The extensive survey from Chandrapur and surrounding including fresh water bodies *viz.*, ponds, streams, field grasslands, and forest areas recorded five families of damselflies with 12 genera. The checklist adds 23 new records for Tadoba national park. Of the total 64 species, 23 were abundant or very common, 24 were common, 12 rare and five very rarely in occurrence. Triple *et al.*, (2012) studied dragonflies and damselflies of tropical forest research institute, Jabalpur, Madhya Pradesh, central India. In the study, 48 species of odonates from two suborders in nine families were observed. In this, 15 were mostly found, 15 were normally found and 16 were rarely found. Tijare and Patil (2012) studied Gorewada national park wherein 29 species of odonates from 24 genera and six families were recorded and *Ceriogrion coromandelianum* was the mostly occurring species in the entire survey from Gorewada, Nagpur. The other species found are *Ceriogrion coromandelianum, Ischnura aurora, Agriocnemis pygmea, Copera marginipes, Lestes viridulus* and *Lestes umbrinus*. The wetland have various Odonata fauna and they have one prominent feature which occurred on the basis of seasonal, perennial, surrounded by trees, shrubs, bushes and grass meadows and this type of habitat usually used by the certain species of the damselflies for breeding purposes. Collection was usually done on every Sundays, at an optimum temperature. Husain *et al.*, (2012) studied new records of dragonflies from Chhatarpur district, Bundelkhand, Madhya Pradesh, India. *Orthetrum pruinosum neglectum, Potamarcha congener, Diplacodes trivialis, Bradinopyga geminata, Pantala flavescens* and *Trithemis aurora* were observed for first time in Chhatarpur district. Manwar *et al.*, (2012) studied the diversity and abundance of odonates of Chatri lake region, from the Amravati, Maharashtra wherein 22 species of odonates from four families were observed. Family Libellulidae shows dominance followed by Coenogrionidae. Lankika *et al.*, (2012) observed odonates diversity in Meegahawatta, a wetland area in Hanwella, Sri Lanka wherein 16 species of anisopterans and 11 species of zygopterans were recorded. *Neurothemis tulia tulia* from Anisoptera *and Pseudagrion malabaricum* from Zygoptera were the abundant species. Species diversity and evenness was lower in Anisoptera than Zygoptera.

Andrew (2013) reported a total of 34 odonates species belonging to the family: Coenagrionidae (seven species), Lestidae (one species.), Aeshnidae (three species.), Gomphidae (one species.) and Libellulidae (22 species) from the Zilpi lake, Maharashtra. Shende and Patil (2013) studied the diversity of dragonflies from Gorewada International Bio-Park, Nagpur, and reported 34 species belonging to 24 genera and four different families. Out of total dragonfly species examined, 26 (76.47%) were common and 8 (23.53%) occasional. Libellulidae family has maximum number of genera and species followed by Aeshnidae,

Gomphidae and Macromiidae. Sahoo *et al.*, (2013) studied the Kanha tiger reserve areas of central India located from the eastern base of the triangular Satpura hilly ranges, which recorded a distribution of 38 species of odonates including seven families and 26 genera, where 12 species were recorded for the first time. The family Libellulidae (21) is well represented in the present study followed by Coenagrionidae (eight species), Calopterygidae (two species), Gomphidae (two species), Lestidae (two species) and Aeshnidae (three species). In the case of zygopterans, *Ischnura aurora* was more abundant than the others. Among the collected libelluids, *Orthretum Sabina* was the most abundant species. Of the 38 species, seven were abundant, 25 common, five occasional and one was rare. Tiple *et al.*, (2013) studied water bodies of the Vidarbha region of central India in the year 2006-2012, wherein 82 species of odonates were recorded. In this investigation 13 new species of odonates were added from Vidarbha region and six species added from Maharashtra where 23 species were abundant or very common (VC), 26 species were common (C), 24 species rare (R) and nine species very rare (VR). Anbalagan *et al.*, (2013) studied odonates diversity in fields of brinjal, okra and rice fields from January 2005 to December 2008 in Thiruvallur district of Tamil Nadu, where 23 species of the dragonflies and 12 species of the damselflies (Zygoptera) belonging to eight families were recorded. Vegetables fields showed higher diversity than rice field during the entire study. Diversity indices clearly showed that maximum odonata diversity was found in vegetables fields than rice fields. Babu *et al.*, (2013) made a checklist on the endemic species which were restricted within the geographic region and their documentation was very much useful in studying the bio-geographical distribution. Endemic species of the odonates exhibited narrow geographic distribution. A total of 503 species and subspecies of Indian odonates wherein 186 species belonging to 67 genera of odonates are endemic to India. Singh and Banyal (2013) studied odonates fauna of Khajjiar lake of Chamba district of Himachal Pradesh, India which recorded 10 species from eight genera and five families of odonates. It has been observed that the family Libellulidae supported the highest number of species (six species, under four genera) and all other families have been represented by a single species each and eight species belonged to Anisoptera and two to Zygoptera.

Dawn and Chandra (2014) studied the Odonata diversity of Chhattisgarh, India, which recorded 68 species from 44 genera and 10 families, where 13 species from 12 genera are observed for the first time in the state. In the entire study family Libellulidae showed dominance with 35 individuals. Sonawane (2014) made a comparative study on Bhima river and Meena river which revealed eight species of the damselfly in collection sites, six species revealed from Bhima River and four species recorded from the Meena River. In the entire investigation two families are reported during the study period, family Coenagrionidae comprises large amount of the population rather than family Platycnemididae. In the investigation *Ceriagrion coromandelianum, Ischnura senegalensis, Pseudagrion rubriceps* and *Copera marginipes* are found in both the region. Bharamal *et al.,* (2014) recorded 23 species from 13 genera and four families from natural habitat of Sindhudurga district. The family Libellulidae (13 species) was observed to be the most abundant in the habitat, Coenagrionidae, Aeshnidae and Corduiidae includes three species each and one species from family Gomphidae. Varghese *et al.,* (2014) studied odonata diversity of Salim Ali bird sanctuary and its adjacent areas in Thattekad, Kerala wherein 51 species of anisopterans and 31 species of zygopterans were recorded of which family Libullilidae showed dominance. Adarsh *et al.,* (2014) made a preliminary checklist of odonates in Kerala Agricultural University (KAU) campus, Thrissur district, Kerala, southern India and reported 36 species of dragonflies and 16 species of damselflies Libellulidae showed dominance in dragonflies and family Coenagrionidae showed dominance in the damselflies. Bora and Meitei (2014) recorded odonates from Indian Council of Agricultural Research (ICAR), Research Complex for NEH Region Campus, Umiam, Meghalaya, India in which 33 species of odonates were collected and the most abundant species was found from Libellulidae with 21 species, followed by two species from family Gomphidae and one species from family Aeshnidae. In Zygoptera nine species were observed from family Coenagrionidae. Sajan *et al.,* (2014) studied diversity and abundance of odonates in Palamau Tiger Reserve, Jharkhand which revealed a total 30 species of odonates belonging to 26 genera of two suborders and eight families. Joshi and Kunte (2014) studied dragonflies and damselflies of Nagaland which showed 68 odonates species from 13 familes and 54 genera. Saha and Gaikwad (2014) observed diversity and abundance of odonates in parks and gardens of Pune city, wherein 1113 individuals from 33 species in six families were observed. Libellulidae showed dominance with 20 species, Coenagrionidae with seven species Platycnemidae and Aeshnidae with two species, Macromiidae and Gomphidae with one species each. Baruah and Saikia (2015) studied abundance and diversity of odonates in different habitats of Barpeta district, Assam, India wherein a total of 45 species of Odonata including 29 species under three families of Anisoptera and 16 species under three families of Zygoptera were recorded in four different types of habitats in Barpeta district of Assam. *Diplacodes trivalis* was dominant in ponds, *Rhyothemis variegate* was dominant in rivers and it was *Pantala flavescens* which was dominant in the open habitat. Das *et al.,* (2012) recorded diversity, distribution and abundance of dragonflies and damselflies in Buffer areas of Similipal Tiger Reserve, India. The entire study reported 58 species whereas 35 species from Anisoptera and 23 species from Zygoptera. The family Libellulidae was found dominant at the study site.

Basumatary *et al.,* (2015) recorded diversity of odonata in Bodoland University, Assam, India which revealed 34 species which includes 26 species of dragonflies (Anisoptera) which belonged to three families and eight species of damselflies (Zygoptera) belonged to three families. Libellulidae from Anisoptera, was the richest family with 20 species and the Coenagrionidae from the Zygoptera was dominant with six species. Adarsh *et al.,* (2015) studied odonates diversity of Chinnar Wildlife Sanctuary, Idukki district, Kerala, which recorded 48 species of odonates, which had 31 species of Anisoptera (Dragonflies) and 17 species of Zygoptera (Damselflies). Among the dragonflies, family Libellulidae dominated with 25 species, followed by Aeshnidae (four species) and Gomphidae (two species). Among Zygoptera, the Coenagrionidae (seven species) was the dominant family followed by Calopterygidae (three species) and Platycenemididae (three species). Das *et al.,* (2015) recorded diversity, distribution and abundance of damselfly (Zygoptera) in wetland of Barpeta district, Assam, India. The study recorded 26 species of Zygopetra from three families and 11 genera. Family Coenogrionidae found to be dominant in the study site. Kalita *et al.,* (2014) studied odonates of Manchabandha reserve forest, Baripada, Orissa which recorded 48 species of odonates comprising of 33 species from Anisoptera and 15 species from Zygoptera. Family Libellulidae showed dominance in Anisopetra and family Coenagrionidae showed dominance among Zygoptera.

Gajbe (2015) studied odonates fauna of Karhandla region of Umred Karhandla Wildlife Sanctuary, Maharashtra, India. The study revealed 28 species of odonates inhabiting the study area which included 22 species of dragonflies and six species of damselflies. Among the

Anisoptera, Libellulidae dominated with 19 species of dragonflies, while family Gomphidae was represented by single species and family Aeshnidae by two species. Among the six species of Zygoptera recorded, five species of damselflies belong to family Coenagrionidae, while family Lestidae is represented by one species. Koli *et al.,* (2015) studied diversity and species composition of odonates in Southern Rajasthan, India wherein a total of 1,290 individuals from eight families and 54 species were recorded, four families and 28 species belonged to Anisoptera, four families and 26 species belonged to Zygoptera. Libellulidae was observed to be the dominant.

Painkra *et al.,* (2016) studied diversity of odonates in gwarighat region of river Narmada, Jabalpur Madhya Pradesh where a total of 22 species of odonates were sampled. Of which nine species were reported from Libellulidae. Most of the organisms were aggregated due to habitat specific nature and random distribution indicates availability of resource utilization to survive but, in the urban forest area high anthropogenic disturbances were noticed that creates huge biotic pressure on forest habitat. The study recorded 22 species from order Odonata which includes 14 dragonflies and eight damselflies. In damselfly the family Coenagrionoidae showed dominance with seven species, then the family Chlorocyphidae with one species and in the dragonfly family Libellulidae has nine species which showed dominance among dragonflies followed by Aeshnidae with three species and Gomphidae with two species. Naik and Sayeswara (2017) studied diversity, occurrence and abundance of odonates of tunga river bank, adjoining fields and cultivated lands in Shivamogga district of Karnataka, India which recorded a total of 29 species which included 14 species of Anisoptera and 15 species of Zygoptera. In Anisoptera, Libellulidae showed dominance and in Zygoptera, Coenagrionidae showed dominance

2.2. Habitat of odonates

The odonates mainly depended on water bodies. They are seen in large range near running and stagnant water. They inhabit all kind of freshwater habitats either temporary or permanent. Most of the species are strictly attached to certain habitats and some of them are also adapted to man-made water bodies. Habitat specificity plays a significant role in the distribution and ecology of odonates. The species of hill streams tend to be narrowly distributed when compared to pool breeders, which are widespread. The adult are closely related with the water bodies and the habitat around them for their oviposition and larva stage of the odonates adapt the aquatic mode of life. This forms a relationship between the species and their habitat (Corbet, 1993, 1999) which affects odonates when there is a change in their habitat (Corbet, 1993; Clark and Samways, 1996; Schindler, Fesl, and Chovanec 2003; Foote and Rice Hornung 2005). Odonates are habitat specific wherein there is a sudden change in the habitat affects them. They are also good indicators of fresh water habitat (Watson *et al.,* 1982; Brown, 1991; Martín and Maynou, 2016).

2.3. Seasonal studies on odonates:

Palot and Soniya (2000) reported 16 species of odonates under two families and 14 genera. In Anisoptera the family Libellulidae represented 11 species while in the Zygoptera the family Coenagrionidae represented five species. In the summer season, a total of 10 species were observed while in the winter twelve species were recorded from the site.

Asiathambi and Manickavasagam (2002) studied odonates of Annamalai University, Annamalaingar, Tamil Nadu, wherein 23 species under four families and 21 genera were collected and identified. Suborder Zygoptera is represented by the family Coenagrionidae and Anisoptera by the families Libellulidae (13 species), Aeshnidae (four species) and Gomphidae (one species). Among the collected Libelluids, *Pantala flavescens* and *Ortherum sabina* were abundant during monsoon season. Among zygopterans *ceriagrion coromandelianum* was the most abundant. Rathod *et al.,* (2012) studied the diversity and abundance of dragonflies and damselflies in agro ecosystems around Amravati city in monsoon season wherein a total 31 species belonging to six families of dragonflies and damselflies were recorded, in which the most abundant family was Libellulidae followed by Coenagrionidae, while Gomphidae, Lestidae, Aeshnidae, Platycnemididae families were the least abundant. Libellulidae family represented 17 species, Coenagrionidae represented nine species, Gomphidae represented two while Aeshnidae, Platycnemididae and Lestidae were with one species each.

Kawade (2013) studied diversity and abundance of damselflies of Saikheda water reservoir of Yavatmal district, Maharashtra which revealed 10 species of damselflies from three families and seven genera eighty percent was observed from family Coenogrionidae while family Platycenemididae and Protoneuridae showed one species. Kulkarni and Subramanian (2013) collected 609 specimens during Pre Monsoon, Post monsoon and winter season in which they reported 46 species of odonates in 26 genera and eight families from Mula Mutha river. The most dominant species was *Pantala flavescens* (Fabricius, 1798) which accounted for 23% of total number of individuals found during sampling. This dominance could be due to mass emergence of the species after monsoon and their yearly aggregation before migration. Individuals of this species congregate every year before migration to Eastern Africa. This was followed by *Brachythemis contaminata* with 17% individuals found only in urban and agricultural landscapes. The third most abundant species was *Trithemis festiva* (Rambur, 1842) with 11% of individuals that were recorded mostly from forested areas except a single record from one garden pond in an urban area. *Trithemis festiva* is known to breed in sluggish streams and usually perches on boulders and aquatic plants.

Govindan *et al.*, (2015) studied distribution of dragonflies and damselflies from Pulicat lagoon, Tamil Nadu wherein 18 species of odonates from three families were recorded. It has been observed that high abundance of odonates was seen in monsoon season. *Orthetrum sabina* showed more dominance followed by *Brachthemis contaminata, Ceriagrion coromandelianum* while *Copera marginipes* showed the least dominance.

Muthukumaravel *et al.*, (2015) studied the seasonal variation of dragonfly diversity in Muthupet mangrove forest, Tamil Nadu which recorded eight species *viz.*, *Rhyothemis variegata, Anax guttatus, Pantala flavescens, Brachythemis contaminata, Orthetrum sabina, Diplacodes trivialis, Crocothemis servilia* and *Tramea basilaris* of Anisoptera which included in two families. Libellulidae showed dominance with seven species and one species in Aeshnidae. Diversity of species was high in monsoon and observed less in the months of summer. Dayakrishna and Arya (2015) revealed diversity and abundance of odonates fauna in the Corbett Tiger Reserve habitat where 420 individuals of odonates from 19 species and four families in two suborders were observed. In Anisoptera, Libellulidae was the only family found from the study area and three families *viz.*, Coenagrionidae, Chlorocyphidae and Calopterygidae were from the Zygoptera. Libellulidae showed dominance with 15 species followed by Coenagrionidae with two species, Calopterygidae and Chlorocyphidae with one species. *Orthetrum pruinosum* was found most in number followed by *Aethriamanta brevipennis, Orthetrum glaucum* and *Crocothemis servilia*. High diversity and abundance of odonates were observed in monsoon season followed by summer and winter.

Kannagi *et al.*, (2016) studied diversity of Anisopetra in a deciduous forest of Thoothukudi district, Tamil Nadu which recorded 958 individuals from 20 species under 16 genera and four families. Libellulidae showed dominance in the habitat with 15 species followed by one species from the family Aeshnidae, Chlorogomphidae and Gomphidae. In the entire study the maximum diversity is observed in post monsoon seasons. Narender *et al.*, (2016) studied diversity and seasonal abundance of odonates in sawanga vithoba lake, India. The diversity was studied in monsoon, winter and summer wherein a total of 33 species in six families were observed. Libellulidae showed dominance with 19 species, Coenagrionidae with 10 species, Gomphidae, Aeshnidae, Platycnemididae and Lestidae with one species each. High diversity was observed in monsoon followed by winter and summer. *Brachythemis contaminata* showed dominance in the monsoon season.

2.4. Biological significance of Odoantes:

Clark and Samways (1996) stated dragonflies are used as indicators to inspect the biotope quality in the Kruger National Park, South Africa. Spencer *et al.*, (1999) studied damselflies as a bio controlling agents against agricultural, horticultural and forest pest. Odonates during larval stages act as important bio controlling agents over mosquito eggs which have been reported by many taxonomists. Corbet (1999) stated damselflies are the augmentative release of bio controlling agents and are used in many countries for the elimination of the mosquitoes. Odonates are found in almost all kind of habitat ranging from permanent lake, running waters and lakes to small temporary rain pools. Some of them are habitat specific and their distributions are very much reported as voracious predators in larval and adult life stages. They feed mainly on living prey. They are with predatory mouth parts (labium) that can be extended to catch the prey. They are predators in the food chain. They are often used as indicators for analyzing healthy status of environment and conservation management (Corbet, 1999). Their sensitivity to habitat specific quality (e.g., forest cover and water chemistry) and amphibious habits make them suitable for analyzing any change in the environment for a constant period and in the short term both above and below the water surface (Clark and Samways, 1996; Clausnitzer, 2003; Osborn, 2005). Grimaldi and Engel (2005) stated the biogeography of order Odonata as a wealthy fauna. Some odonates showed their range of occurrence to be influenced with environmental changes and northward range extension have been revealed over the last 40 years in several European taxa.

Ghahari *et al.*, (2009) studied odonates from northern Iran, with comments on their presence in rice fields. Odonata plays a vital role in controlling pests in rice fields. Sanchez-Herrera and Ware (2010) reported that odonates are useful in the studies of ecology, behavior, evolutionary biology and biogeography of different places. Damselflies which showed conspicuous behavior, attractive colours and small number of species made a great impact on odonatological study. Damselfies occupies a top position in freshwater ecosystems as a predator. Adarsh *et al.*, (2014) observed damselflies which can be used as indicators of the environment conditions which includes the aquatic and forest habitat. The damselflies can be widely used as indicators to indicate the status of environment in most parts of the world. In recent years, research in the field of ecological impacts has turned increasingly due to the bioindicator organisms because of their specific and known responses to changes in environmental conditions (Godoy *et al.*, 2019, Molineri *et al.*, 2020). Odonata are an amphibious insect order which are widely used as ecological indicators of freshwater ecosystem health (O'Malley *et al.*, 2020).

3.0. AIM AND OBJECTIVES OF THE PRESENT STUDY

- To study the distribution and diversity of odonates in Tiruvallur district, Tamil Nadu, India.
- To measure their diversity indices.
- To study the seasonal availability of odonates in Tiruvallur district, Tamil Nadu, India.
- To study the different kind of diversity profiles of the various habitat and to suggest conservation strategies.

4.0. MATERIALS AND METHOD

4.1. Study area:

Tiruvallur district (12°15' and 13°15' N; 79°15' and 80°20' E) an administrative district of Tamil Nadu, South India comprises of both urban and rural characteristics. It covers an area of about 3422 sq. km. The district receives a moderate rainfall of about 1104mm, (52% of rain by north east monsoon and 41% by south west monsoon). It is an industrial district has 12 taluks, five municipalities, 10 town panchayats, 14 panchayat unions, 526 village panchayats, 54 firkas and 825 revenue villages. Red non-calcareous and coastal alluvial soils are mostly seen and sandy oil with soda or alkali are also observed. Erinaceous kinds of soil are seen in the coastal region. The district mainly depends on the agriculture and its associated activity. The main crops cultivated in the district are rice, groundnut, ragi, cumbu, green gram, black gram, and sugar cane.

The present studies of odonates were carried from January 2015- December 2016. The present study was conducted to survey, photograph and to collect the odonata fauna from various parts of Tiruvallur district, Tamil Nadu, India (Plate: 1). This study was focused on the nearby aquatic bodies. From Tiruvallur district seven study sites were selected for the study of diversity and seasonality of odonates (Plate: 2).

4.1.1. Poondi lake:

Poondi reservoir or Sathyamoorthy Sagar built up in the year 1944 is located at Poondi, a small village (13° 11' 6" N and 79°51' 36" E). The reservoir was built across the Kosathalaiyar river in district of Tiruvallur, Tamil Nadu which as a quantity of 2573 Mcft used to store Kosathalaiyar river water. The reservoir is spread around 121/2 square miles. The water from this reservoir is supplied to Red Hills and then sent to Chennai city. The habitat around the reservoir supports the rich vegetation. Herbs, shrubs and emergent littoral vegetation are observed in huge quantity around the habitat. They support a wide range of habitat with dominated green vegetation and water habitat doesn't suffer from pollution causing activities.

4.1.2. Puzhal lake:

Puzhal lake (13°10' 00" N and 80°10' 17.5" E) also known as the Red Hills lake, is located in Red Hills, Tiruvallur district, Tamil Nadu. It plays a vital role in supplying water to the Chennai city. The full capacity of the lake is 3,300 million ft^3 (93 million m^3) and its surface

19

area is 4500 acres. The catchment areas are present to its north, west and south. The northern and western parts of the catchment area are covered by agricultural land and residential area, where the southern part is surrounded by residential areas. The Redhills tank supplies water through closed conduits to Kilpauk water works for treatment and supply to North Chennai city. The habitat around the Puzhal lake supports good natural vegetation.

4.1.3. Sholavaram lake:

Sholavaram lake is located in Sholvaram village of Tiruvallur district, Tamil Nadu (13.22757 ° N and 80.15024 ° E) is one of the important reservoirs from where water is supplied to Chennai city. The western side of the catchment area of the lake is covered by agricultural land and residential areas, where the south part of the catchment area includes agricultural land, residences and small industries. The vegetation around the lake includes a rich source of herbaceous vegetation around the habitat which supports a rich diversity of many organisms.

4.1.4. Avadi lake:

Avadi lake is located in the locality of Avadi a municipality in Tiruvallur district, Tamil Nadu (13.12°N 80.1°E). This lake is of 2.64 km in length, and is surrounded by residential areas and industries. The vegetation is widely spread but in recent days the human settlements around the lakes have caused disturbances in vegetation and some part of the lake has been covered by water hyacinths.

4.1.5. Ambattur lake:

Ambattur lake, is a rain-fed reservoir which is located at Ambattur taluk of Tiruvallur district, Tamil Nadu (13.10°N 80.14°E). Ambattur lake once served as a main water resource for the surrounding areas, but now has been polluted. The lake is around 250 acres, Ambattur is the largest industrial estate and surrounded by many industries. The lake also faces the problems like garbage dumping, water mining, construction of buildings and industries and pollution from the sewage.

4.1.6. Korattur Lake:

Korattur lake is a chain of three lakes comprising of Ambattur lake, Madhavaram lake and Korattur lake, spread over 990 acres in Korattur, Tiruvallur district, Tamil Nadu (13°07'19.2"N 80°11'04.4"E). The surface area of the lake is 990 acres. The lake has a moderate source of vegetation and it found to be polluted and suffering from eutrophication. Due to rapid population growth and anthropogenic activities the survival of this lake has become miserable.

4.1.7. Poonamallee:

Poonamallee (13.05°N 80.11°E) is a town in the Poonamallee taluk of the Tiruvallur district in Tamil Nadu. The area is presently considered to be a suburb of Chennai. It has been developed rapidly and natural vegetation in the areas are disturbed by human activities. It is surrounded by residential houses and industries on all the sides and herbaceous vegetation is moderate around this habitat.

Regular collection and observation was made in the seven study site of Tiruvallur district, Tamil Nadu, India during early morning, midday and late evening hours on weekly basis. Diversity of odonates were studied by using line transects method (Moore and Corbet, 1990; Brooks, 1993). For sampling, sweep net technique was used with a 30-45cm (dia) and a depth of 80cm hand net. At a fixed transect of 500m length, observations were made. Observations were made through walking in a wide area of the site with the aid of binocular and digital cameras. The surveys of adult odonates were carried out from the month of January 2015 to December 2016 randomly in morning and evening in good weather conditions. The study has been carried out by visiting the site once in a week. Adult odonates were sampled for a period of two years (2015 and 2016) in seven habitats across the study area. Sampling of adult odonates were done on days with fine weather conditions in all the seasons and in all the habitats by moving along transects. During the study period, the diameter of the net and the total number of sweeps made were kept constant where parameters, *viz.*, pH, temperature and rainfall were also observed. The relevant meteorological data including air temperature, rainfall and relative humidity have been collected from the nearest Regional Meteorological Station, Nungambakkam Chennai. The estimation of population density was done and data compiled. Species caught by sweep net were photographed for identification. Species were identified with the help of identification keys (Fraser 1931, 1933, 1934, 1936), Emiliyamma (2005), Subramanian (2005, 2008) and Andrew *et al.,* (2009). The abundance was determined by the number of individual encountered during visual surveys along transects. The total number of individuals of each species encountered during the two years of sampling period (January, 2015 to December, 2016) was used for comparing their abundance. In the present study, the diversity of odonates around seven habitats were studied. Population tendency were observed in all the sites following line transect counting technique. In all the study sites, the same method of survey was carried out.

The different types of habitat which surveyed were natural and those areas with human interference. Sites like agricultural and farm lands which had no permanent water body like ponds, or river were also surveyed. Most of these areas were open surrounded by shrubs and trees. Paddy fields, vegetable fields or lands with plantation were also surveyed. Few open land stretches which had water filled in rainy season acted as breeding sites for most of the odonate species. The families in the odonate species were classified on the basis of their abundance in the study sites. To study the odonata biodiversity, the odonates were observed and photographed in their natural environment and a few individuals were collected and photographed in high magnification to know their classification. The wing venation, colour pattern and genitalia were observed for identification purposes. Details such as collection date and locality of all specimens were recorded. Simpson, Shannon-Wiener, and Margalef indices were used to measure the diversity of odonates in different study sites (Shannon-Wiener, 1949; Simpson, 1949; Margalef, 1958).

5.0. RESULTS

5.1. Diversity of odonates in the seven habitats of Tiruvallur district

In the entire study, a total of 22 species of odonates were identified in all the study sites. The dragonflies showed dominance over damselflies in all the habitats.

Fourteen species of Anisoptera under three families *viz.,* Libellulidae, Aeshnidae and Gomphidae were observed in the study (Table: 1). Libellulidae showed dominance with ten genera and twelve species (Plate: 3) while Aeshnidae (Plate: 4) and Gomphidae (Plate: 5) were represented by one species each. The observed dragonflies in the study sites includes *Brachythemis contaminata, Crocothemis servilia, Diplacodes trivialis, Neurothemis tullia, Orthetrum pruinosum, Pantala flavescens, Rhyothemis variegata , Trithemis aurora, Tholymis tillarga, Tramea limbata, Orthetrum glaucum, Orthetrum sabina, Anax guttatus* and *Ictinogomphus rapax* The diversity of dragonflies from all the study habitats in the year 2015 and 2016 were studied (Figure: 1 and Figure: 2).

Eight species of Zygoptera under three families *viz.,* Coenagrionidae, Calopterygidae and Platycnemididae were observed in the study (Table: 2). In this family Coenagrionidae (Plate: 6) showed dominance with four genera and five species, followed by family Calopterygidae (Plate: 7) with one genera and two species and one species from family Platycnemididae (Plate: 8). The observed damselflies in the study sites includes *Agriocnemis pygmaea, Ceriagrion cerinorubellum, Ceriagrion coromandelianum, Ischnura aurora, Pseudagrion microcephalum, Vestalis apicalis, Vestalis gracilis* and *Copera marginipes*. The diversity of damselflies from all the study habitats in the year 2015 and 2016 were studied (Figure: 3 and Figure: 4).

In the two year study of odonata diversity from the Poondi habitat in the year 2015 a total of 1433 individuals of dragonflies and 1623 individuals in the year 2015 were recorded from 14 species belonging to three families. Libellulidae showed dominance in all habitats. In both the years *Diplacodes trivialis* was abundant in the study site followed by *Pantala flavescens and Crocothemis servilia. Anax gutatus* showed less abundance in both years. In the case of damselflies, during the year 2015, a total of 668 individuals were recorded from eight species in three families while in the year 2016, it was a total of 583 individuals from eight species in three families. Coenagrionidae were the dominant family where the *Ceriagrion*

23

coromandelianum showed more dominance followed by *Pseudagrion microcephalum, Ceriagrion cerinorubellum* and *Agriocnemis pygmaea* while *Ischnura aurora* showed low abundance in both the years.

From the habitat of Puzhal lake in the year 2015, a total of 1392 individuals and in the year 2016 it was a total of 1242 individuals of dragonflies from 14 species in three families. Libellulidae showed the highest dominance in the habitat. Among the dragonflies, *Diplacodes trivialis* showed dominance in both the years followed by *Pantala flavescens* and *Orthetrum sabina* where *Anax gutatus* showed the least dominance. In the case of damselflies in the year 2015, a total of 583 and in 2016 a total of 514 individuals from eight species in three families were recorded. In both the years, among the damselflies Coenagrionidae was the dominant family where *Ceriagrion coromandelianum* showed the highest dominance followed by *Pseudagrion microcephalum* and *Ceriagrion cerinorubellum* while *Vestalis apicalis* was the least dominant species.

From the habitat around the Sholavaram in the year 2015, a total of 1406 individuals observed from 14 species in three families and in 2016, a total 1250 individuals recorded from 14 species belonging to three families were recorded. Libellulidae showed the highest dominance in the habitat while *Diplacodes trivialis* showed more dominance in both the years followed by *Pantala flavescens* and *Orthetrum sabina. Anax gutatus* showed least dominance in both the years. In the year 2015, a total 485 individuals of damselflies were recorded and in 2016, a total of 417 individuals from eight species in three families were observed. In damselflies, Coenagrionidae was the dominant family where *Ceriagrion coromandelianum* showed abundance followed by *Pseudagrion microcephalum*. The species *Vestalis apicalis* and *Ischnura aurora* showed less abundance in both the years.

From the two year study of diversity of odonates in the Avadi habitat during the year 2015, a total of 1017 individuals of dragonflies were recorded from 13 species in two families and in 2016, a total 868 individuals were recorded from 13 species in two families In both the years, Libellulidae showed dominance in the habitat while *Diplacodes trivialis* was the most abundant species followed by *Orthetrum sabina* and *Crocothemis servilia* whereas *Tholymis tillarga* and *Ictinogomphus rapax* were found less in abundance. In 2015, a total of 465 individuals of damselflies were observed while in 2016, a total of 427 individuals were recorded from seven species in three families, where Coenagrionidae was the abundant family. In the entire study, *Ceriagrion cerinorubellum* was abundant in both the years followed by *Ceriagrion coromandelianum* and *Copera marginipes* whereas *Vestalis apicalis* was the least abundant species.

From the habitat around the Ambattur lake in the year 2015, a total of 1223 individuals of dragonflies and in 2016, a total 1217 individuals were recorded from 13 species in two families. Libellulidae showed the highest dominance in the habitat in both the years while *Diplacodes trivialis* was the most abundant species followed by *Orthetrum sabina, Brachythemis contaminata* and *Crocothemis servilia. Tholymis tillarga* and *lctinogomphus rapax* were the least abundant species. In damselflies, a total of 548 individuals were recorded in the year 2015 and a total of 494 individuals in 2016 from seven species in three families. Coenagrionidae was the abundant family in the study and in 2015 *Ceriagrion cerinorubellum* was dominant followed by *Copera marginipes* and *Ceriagrion coromandelianum.* Whereas in 2016, *Pseudagrion microcephalum* showed dominance followed by *Copera marginipes.* The least dominance was observed from the species *Vestalis apicalis* and *Vestalis gracilis.*

In the habitat around the Koratur lake, during 2015, a total of 902 and in 2016, a total of 779 individuals of dragonflies were recorded from 13 species belonging to two families. Libellulidae was dominant in the habitat. The most dominant species observed *Orthetrum sabina* followed by *Diplacodes trivialis* in both the years where as *Tholymis tillarga* followed by *lctinogomphus rapax* showed less abundance. From damselflies a total of 408 individuals were recorded in 2015 and 392 individual in 2016 from seven species belonging to two families. In both the years, Coenagrionidae showed dominance where *Ceriagrion cerinorubellum* was the abundant species followed by *Ceriagrion coromandelianum* and *Pseudagrion microcephalum. Vestalis apicalis* and *Vestalis gracilis* were the less dominant species.

In the habitats around the Poonamallee from 2015, a total of 1022 and in the year 2016, a total of 787 individuals of dragonflies were recorded from 12 species in a single family. Libellulidae was the only family found in the habitat where *Diplacodes trivialis* showed the highest dominance followed by *Crocothemis servilia* whereas *Tholymis tillarga* showed the least abundance in both the years. In damselflies, from the year 2015, a total of 409 individuals and 335 individuals in 2016 were observed from six species belonging to three families. Coenagrionidae showed the highest dominance in 2015 and the abundant species was *Pseudagrion microcephalum* followed by *Ceriagrion cerinorubellum* and *Ceriagrion coromandelianum.* In 2016, *Ceriagrion cerinorubellum* showed dominance followed by *Pseudagrion microcephalum.* In both the years *Vestalis apicalis* and *Vestalis gracilis exhibited the least dominance.*

5.2. Diversity indices

5.2.1. Shannon Weiner Diversity

The Shannon weiner diversity values of the dragonflies for the year 2015 in the seven study habitat *viz.,* Poondi, Puzhal, Sholavaram, Avadi, Ambattur, Korattur and Poonamallee were 3.725, 3.698, 3.675, 3.609, 3.635, 3.557 and 3.494 (Table: 3) and in 2016, the values were 3.711, 3.679, 3.675, 3.585, 3.654, 3.557 and 3.461 respectively (Table: 4)

The Shannon weiner diversity values of the damselflies for the year 2015 in the seven study habitat *viz.,* Poondi, Puzhal, Sholavaram, Avadi, Ambattur, Korattur and Poonamallee were 2.961, 2.965, 2.956, 2.783, 2.797, 2.783 and 2.556 (Table: 5) and in 2016 the values were 2.943, 2.973, 2.953, 2.775, 2.791, 2.766 and 2.565 respectively (Table: 6).

5.2.2. Simpson Index

The Simpson index diversity values of the dragonflies for the year 2015 in the seven study habitat *viz.,* Poondi, Puzhal, Sholavaram, Avadi, Ambattur, Korattur and Poonamallee were 0.077, 0.080, 0.082, 0.085, 0.082, 0.090 and 0.091 (Table: 3) and in 2016 the values were 0.079, 0.081, 0.082, 0.087, 0.087, 0.090 and 0.095 respectively (Table: 4).

The Simpson index diversity values of the damselflies for the year 2015 in the seven study habitat *viz.,* Poondi, Puzhal, Sholavaram, Avadi, Ambattur, Korattur and Poonamallee were 0.130, 0.129, 0.130, 0.145, 0.143, 0.145 and 0.171 (Table: 5) and in 2016 the values were 0.132, 0.127, 0.130, 0.146, 0.144, 0.148 and 0.168 respectively (Table: 6).

5.2.3. Margalef Richness Index

The Margalef richness index diversity values of the dragonflies for the year 2015 in the seven study habitat *viz.,* Poondi, Puzhal, Sholavaram, Avadi, Ambattur, Korattur and Poonamallee were 1.789, 1.796, 1.823, 1.733, 1.683, 1.801 and 1.589 (Table: 3) and in 2016 the values were 1.759, 1.825, 1.823, 1.774, 1.689, 1.801 and 1.651 respectively (Table: 4).

The Margalef richness index diversity values of the damselflies for the year 2015 in the seven study habitat *viz.,* Poondi, Puzhal, Sholavaram, Avadi, Ambattur, Korattur and Poonamallee were 1.076, 1.099, 1.132, 0.959, 0.951, 0.981 and 0.083 (Table: 5) and in 2016 the values were 1.099, 1.121, 1.160, 0.990, 0.967, 1.005 and 0.860 respectively (Table: 6).

5.2.4. Equitability Index

The Equitability index diversity values of the dragonflies for the year 2015 in the seven study habitat *viz.*, Poondi, Puzhal, Sholavaram, Avadi, Ambattur, Korattur and Poonamallee were 0.978, 0.971, 0.965, 0.975, 0.982, 0.961 and 0.974 (Table: 3) and in 2016 the values were 0.974, 0.966, 0.965, 0.968, 0.987, 0.961 and 0.965 respectively (Table: 4).

The Equitability index diversity values of the damselflies for the year 2015 in the seven study habitat *viz.*, Poondi, Puzhal, Sholavaram, Avadi, Ambattur, Korattur and Poonamallee were 0.987, 0.988, 0.985, 0.991, 0.996, 0.991 and 0.988 (Table: 5) and in 2016 the values were 0.981, 0.991, 0.984, 0.988, 0.993, 0.985 and 0.992 respectively (Table: 6).

Almost in most of the habitats, the high range of diversity was observed in the year 2015 while 2016 showed a little decrease in the diversity.

5.3. Systematic account on the dragonflies (Anisoptera) observed in the study habitat of Tiruvallur district, Tamil Nadu.

        Order: Odonata
        Suborder: Anisoptera
        Super Family: Libelluloidea
        Family: Libellulidae
        Subfamily: Sympetrinae

Genus: *Brachythemis* (Brauer, 1868)

1. Species: *Brachythemis contaminata* (Fabricius, 1793)

Common name: Ditch Jewel

Description: In male the eyes are olivaceous brown above and they are bluish grey beneath, Thorax are olivaceous brown to reddish brown above and also has two reddish brown lateral stripes. The wings are transparent with reddish venation and it also has a deep orange patch spreading from base of the wing is seen in fore wing and hind wing. The abdomen is reddish which has dorsal and sub dorsal brown stripes. In female, the eyes are pale brown in upper and in the lower region they are bluish grey. Thorax region is pale greenish-yellow. The wings are transparent and an orange patch was observed in the male. The hind wing is lightly stained with yellow colour at the base. Abdomen is pale olivaceous brown with black stripes at the middle.

Size: The abdomen of the male is about 18 to 21mm and the hind wing ranges from 20 to 23mm. In female the abdomen is about 18 to 20mm and the hind wing ranges from 22 to 25mm.

Habits and habitat: It is a common species which are seen in large numbers and they usually fly over the water surface. They are seen near freshwater, ponds, tanks and they are also seen near polluted waters.

Distribution: Found in various parts of Tiruvallur district.

Genus: *Crocothemis* (Brauer, 1868)

2. Species: ***Crocothemis servilia*** (Drury, 1770)

Common name: Ruddy Marsh Skimmer

Description: In male, the eyes are blood red colour in the upper region and they are purplish in the sides, thorax range from blood red to bright orange. Wings are transparent and the bases of the wings are striking with yellow colour. Abdomen is blood red which has a blackish colour in mid dorsal carina of nine and tenth segment. In female, the eyes are brown in upper and in lower its olivaceous. Thorax is dark brown in colour. Wings are transparent and the bases of the wings are with pale yellow colour. Abdomen is yellowish brown which has a mid-dorsal black stripe.

Size: The abdomen of the male is about 24 to 35mm and the hind wing ranges from 27 to 38mm. In female the abdomen is about 25 to 32mm and the hind wing is from 31 to 37mm.

Habits and habitat: They are commonly found near all the aquatic habitats, *viz.,* ponds, puddles, rivers, big wells, tanks, ditches, agricultural fields and even in the temporary water bodies.

Distribution: Found in various parts of Tiruvallur district..

Genus: *Diplacodes* (Kirby, 1889)

3. Species: ***Diplacodes trivialis*** (Rambur, 1842)

Common name: Ground Skimmer

Description: They are usually small sized dragonflies generally greenish yellow or blue with black markings. In males, the eyes are reddish brown in the upper and pale blue or yellow colour in lower region. Thorax is generally olivaceous or they will be in greenish yellow. Wings are transparent which has short pterostigma dark reddish brown between black nervures. The abdomen was greenish-yellow with sutures which are finely black. In matured ones, the thorax and abdomen are pruinosed blue. Female generally resembles the male where the abdominal markings are broad and continues to eighth and nineth segment.

Size: The abdomen of the male is about 19 to 22mm and the hind wing ranges from 22 to 23mm. In female the abdomen is about 18 to 20mm and the hind wing is from 22 to 24mm.

Habits and Habitat: They are usually seen near ponds, rivers, slow running streams and agricultural fields but most generally they are seen on roadways a distance off from their habitat. Sometimes they are quite among the soils which are difficult to identify.

Distribution: Found in various parts of Tiruvallur district.

Genus: *Neurothemis* (Brauer, 1867)

4. Species: *Neurothemis tullia* (Drury, 1773)

Common name: Pied Paddy Skimmer

Description: In male, the head region is black in colour and the eyes are blackish brown in the upper and are violaceous. Thorax region is black and they have stripes in the mid dorsal region. They have black colour legs. The apex half region of the wings are transparent while the half of the base are bluish black with a patch of milky white close to the tip. Abdomen is black which has a broad mid dorsal with creamy white stripe in the anterior region. Anal appendages are creamy white in colour and at the edge they are black. Female shows certain differences to the male in colour and body markings. The upper region of the eyes are in faded brown colour and in the side region it was olivaceous. Thorax is greenish yellow and the legs are yellow in colour at the outer while they are black in inner. Basal part of the wing resembles a amber yellow colour while the anal part has blackish brown patch at the edge. Abdomen are bright yellow which has a stripe in black colour. Anal appendages are short which are bright yellow in colour.

Size: The abdomen of the male is about 16 to 20mm and the hind wing ranges from 19 to 23mm. In female the abdomen is about 16 to 19mm and the hind wing is from 20 to 23mm.

Habits and habitat: They are weak fliers usually they are observed near fresh water ponds, semi saline ponds, lakes and in the agricultural fields. They are very commonly seen in the paddy fields.

Distribution: Found in various parts of Tiruvallur district.

Subfamily: Libellulinae

Genus: *Orthetrum* (Newman, 1833)

5. Species: *Orthetrum pruinosum* (Burmeister, 1839)

Common name: Crimson tailed marsh hawk

Description: In male the eyes are blue black in the upper part and bluish grey in the lower part. Thorax is reddish brown to dim brown colour and they are enclosed with fine hairs. Legs are black and reddish brown in colour. Wings are transparent and in matured adult they are pale brown nearing the edge while the base area showed reddish brown markings. Abdomen is purplish red in adults. In female the eyes are yellowish cover with brown colour. Thorax is dull ochreous brown with brown stripes. Wings are identical as that of male. Abdomen is ochreous brown with all the segments lined with black colour.

Size: The abdomen of the male is about 28 to 31mm and the hind wing ranges from 32 to 36mm. In female, the abdomen is about 30mm and the hind wing is 37mm.

Habits and habitat: It is one of the common dragonflies found near the ponds, rivers, wells and ditches. They are usually observed while perching on plants, stones etc.

Distribution: Found in various parts of Tiruvallur district.

6. Species: *Orthetrum glaucum* (Brauer, 1865)

Common name: Blue Marsh Hawk

Description: In male the eyes are deep green covered with reddish brown. Thorax is deep blue or black in colour and may also have fine black or blue black colour. The legs are black in colour. Wings are transparent which has amber yellow in the base while in mature adults the smoky brown wings are seen. First three segments of abdomen are bulged and they comprises of pale blue and black colour. In female, the thorax is olivaceous in the upper and surrounded by huge reddish brown stripes. Legs are in black and yellow colour. Wings are same as that of in male. Abdomen is brown reddish which has a yellow greenish stripe in the mid dorsal.

Size: The abdomen of the male is about 29 to 35mm and the hind wing ranges from 33 to 40mm. In female, the abdomen is about 28 to 32mm and the hind wing from 32 to 37mm.

Habits and habitat: They are widely observed near marshes, streams, rivers and canals.

Distribution: Found in various parts of Tiruvallur district.

7. Species: *Orthetrum sabina* (Drury, 1770)

Common name: Green Marsh hawk

Description: In male the eyes are greenish black. The thorax region is greenish yellow with black stripes. The legs are black in colour. Wings are transparent where the inner edge of the hind wing has yellow colouration. Abdomen is greenish yellow in which the first three segments are green with huge black bands which are tremendously swollen at the edge. Female is identical to that of male but has a thinner abdomen tip.

Size: The abdomen of the male is about 30 to 36mm and the hind wing ranges from 30 to 36mm. In female the abdomen is about 32 to 35mm and the hind wing from 31 to 35mm.

Habits and habitat: It is a common dragonfly found near the agricultural fields, lakes, ponds and gardens. They perch motionlessly on the plants for a high time and are observed at a distance from the water bodies.

Distribution: Found in various parts of Tiruvallur district.

Subfamily: Trameinae

Genus: *Pantala* (Hagen, 1861)

8. Species: *Pantala flaveseens* (Fabricius, 1798)

Common name: Wandering glider

Description: In males the eyes are reddish brown in upper and bluish grey on lower and sides. Thorax is olivaeceous or rusty and covered thickly with yellowish hair. Legs are black and the wings are transparent. The base of the hind wing are dim golden yellow in colour as far as distal that of anal loop which has a narrow apical brown spot limited to the posterior boundary of the wing. Abdomen is bright red with brown colour which are stained with the brick red colour. Segments 1-4 are pale yellow while segments 8-10 are with defined black colour. The Sides of segments 1 to 4 are pale yellow in colour and segments 8 to 10 with sharply-defined black mid-dorsal spots with narrow end of spots at posterior end of first segments. Female is similar to that of the male while eyes are olivaceous brown in the upper part and wings are smoky with no apical brown spot.

Size: The abdomen of the male is about 29 to 35mm and the hind wing ranges from 38 to 40mm. In female the abdomen is about 30 to 33mm and the hind wing is 39 to 41mm.

Habits and habitat: They are widely seen near the agricultural fields, irrigation tunnel, small water bodies, lakes, rivers and also by the roadsides. High swarms are observed in morning and in the evenings. It has been observed during the monsoon winds.

Distribution: Found in various parts of Tiruvallur district.

Genus: *Rhyothemis* (Hagen, 1867)

9. Species: ***Rhyothemis variegata*** (Linnaeus, 1763)

Common name: Common picture wing

Description: They are dark metallic dragonflies. In males the eyes are dark reddish brown in colour in the upper region. Thorax is black with dull metallic green colour and the legs are black. The wings are transparent with golden yellow colour. In the tip of the fore wing, the edge has presence of deep coffee brown coloured spots. The basal region of the wing has a brown colour patch. Abdomen is black with slim and long appendages. In female head, thorax and abdomen are identical to that of male. Wings show certain differences where the edge of the wings are transparent while the basal half are black with brown broad markings. The hind wing is brown gloomy and lengthy towards the tip of the wing, and also has a large yellow patch and yellow spots approaching the edge of the wing.

Size: The abdomen of the male is about 23 to 25mm and the hind wing ranges from 33 to 36mm. In female the abdomen is about 20 to 22mm and the hind wing is 28 to 37mm.

Habits and habitat: They are observed in marshes, agricultural fields, lakes and ponds. They are generally weak fliers and regularly perch on aquatic plants. They are generally seen near aquatic habitats and are very rare to be observed away from the water bodies.

Distribution: Found in various parts of Tiruvallur district.

Genus: *Tholymis* (Hagen, 1861)

10. Species: ***Tholymis tillarga*** (Fabricius, 1798)

Common name: Coral-tailed Cloud Wing

Description: In males, eyes are brown covered with reddish olivaceous in the lower region. Thorax is reddish above with golden yellow colour. Legs are ochreous in colour. Wings are transparent and have broad golden brown colour patch in hind wing and white spot in the center. Abdomen is bright rusty-red and the anal appendages are reddish. Female are similar to that of males. Head and thorax are olivaceous. Wings are transparent and the hind wing lacks white spot while the brown colour patch is dull. Abdomen is olivaceous brown.

Size: The abdomen of the male is about 28 to 33mm and the hind wing ranges from 33 to 37mm. In female the abdomen is about 27 to 31mm and the hind wing is from 31 to 37mm.

Habits and habitat: It is crepuscular dragonfly which is actively seen during the sunset. Generally they are attracted to the light in the night times. It is a quick flier and are commonly observed near ponds, lakes, marshes and in plains.

Distribution: Found in various parts of Tiruvallur district.

Genus: *Tramea* (Hagen, 1861)

11. Species: *Tramea limbata* (Desjardins, 1832)

Common name: Black Marsh Trotter

Description: In males the eyes are dark brown in the upper region while the lower and sides are olivaceous. Thorax is olivaceous which has reddish spreading. They have black legs and the edge of the legs are reddish brown. Wings are transparent and the venations are red approaching the posterior edge of the wing and also has intense irregular markings which are blackish brown. The abdomen is blood red colour and has black markings at the last three segments. Female is identical to that of male but the abdomen has a wide range of black markings.

Size: The abdomen of the male is about 33 to 35.5mm and the hind wing ranges from 44 to 46mm. In female the abdomen is 32mm and the hind wing is from 43 to 46mm.

Habits and habitat: They are seen near lakes, rivers and ponds. They are usually observed in noon times roaming around the water bodies.

Distribution: Found in various parts of Tiruvallur district.

Subfamily: Trithemistinae

Genus: *Trithemis* (Brauer, 1868)

12. Species: *Trithemis aurora* (Burmeister, 1839)

Common name: Crimson Marsh Glider

Description: In male the eyes are crimson colour in the upper and they are in brown colour on their sides. The thorax and abdomen region are violet with crimson. Legs are black in colour. Wings are transparent which has venation in crimson, in posterior part of wings there is a huge amber colour patch. In female the eyes purplish brown in upper and in the lower region they are grey coloured. Thorax is olivaceous in colour which has brown and black stripes. The legs are deep grey in colour with yellow stripes. Abdomen is ochreous which has black colour markings. Wings are transparent and the venation varies from bright yellow to brown colour but the posterior amber patch is pale.

Size: The abdomen of the male is about 21 to 29mm and the hind wing ranges from 24 to 34mm. In female the abdomen is about 19 to 27mm and the hind wing is from 24 to 31mm.

Habits and habitat: It is one of the common dragonflies found near aquatic habitats and streams. The adults are usually observed while perching on the aquatic plants and dry twigs near the aquatic habitat.

Distribution: Tiruvallur district.

Superfamily: Aeshnoidea

Family: Gomphidae

Subfamily: Aeshninae

Genus: *Anax* (Leach, 1815)

13. Species: *Anax guttatus* (Burmeister, 1839)

Common name: Blue tailed green darner

Description: In males the eyes are blue along with yellow and in lower they are black. Thorax is pale green in colour, black colour legs are seen and the femora is yellow in colour. Wings are transparent and in the hind wing a huge patch which is amber yellow is seen. In abdomen, the first and second segment are pale green and the second segment is turquoise blue at the back and the segment 4-7 has three pairs of orange bright spots. The female is identical to that of the male but in hind wings there is absence of amber patch. In abdomen the spots which are in orange colour are highly confluent.

Size: The abdomen of the male is about 56 to 62mm and the hind wing ranges from 50 to 54mm. In female the abdomen is about 56 to 58mm and the hind wing range from 52 to 54mm.

Habit and habitat: They are usually observed in noon time and are rarely seen in night light. They are very common near ponds, lakes, rivers and marshes and seen along the water side habitat. The adults are strong in swift flight.

Distribution: Found in various parts of Tiruvallur district.

Subfamily: Lindeniinae

Genus: *Ictinogomphus* (Cowley, 1934)

14. Species: *Ictinogomphus rapax* (Rambur, 1842)

Common name: Common club tail

Description: In male the eyes are bluish grey in colour. Thorax is black in colour which has yellow markings where the back side of the thorax has a huge yellow central spot and yellowish green stripes. The legs are black in colour and the yellow stripe is seen in the inner side of the forelegs. Wings are transparent and they are lightly enfumed in mature adults. Abdomen is black which has yellow markings. The eighth segment of the abdomen is highly dilated and has a wing like projections. Female is identical to that of male and it has wide ranging yellow markings. Abdomen is stouter and short.

Size: The abdomen of the male is 52mm and the hind wing is 40mm. In female the abdomen is about 50mm and the hind wing is from 42 to 44mm.

Habits and habitat: They are commonly observed near ponds, lakes, rivers and tanks. They are fast fliers and been observed as a distinct sitting position in the twigs around the water body by head downwards and the abdomen upwards. Whenever they are disturbed they travel towards the water surface, then swift off and travel around the border of the habitat returning back to the original resting place.

Distribution: Found in various parts of Tiruvallur district.

5.4. Systematic account on the damselflies (Zygoptera) observed in the study habitat of Tiruvallur district, Tamil Nadu.

Order:	Odonata
Suborder:	Zygoptera
Super Family:	Coenagrionoidea
Family:	Coenagrionidae
Subfamily:	Agriocnemidinae

Genus: *Agriocnemis* (Selys, 1877)

15. Species: *Agriocnemis pygmaea* (Rambur, 1842)

Common name: Pigmy dartlet

Description: In males the eyes are black in the upper region and are pale apple green in the lower side. The upper side of the thorax is black and the sides have pale apple green colour stripes. Legs are yellow in colour and the femora is black. Wings are transparent and the fore wing is pale yellow while the hind wing is black. Upper region of the abdomen is black while the segments one to six are pale apple green and the segments at the end are brick red in colour. Females are tough than males with certain differences in the colour forms where the head thorax and abdomen exhibits brick red colour.

Size: The abdomen of the male is about 16 to 17mm and the hind wing ranges from 9.5 to 10mm. In female the abdomen is 18mm and the hind wing ranges from 11 to 12mm.

Habits and habitat: They are usually seen near marshes, rivers, ponds and lakes. They are found in the midst vegetation and flies close to ground. They perch on little grass blades, bounding on the small midges and seat on the stem.

Distribution: Found in various parts of Tiruvallur district.

Subfamily: Pseudagrioninae

Genus: *Ceriagrion,* (Selys, 1876)

16. Species: *Ceriagrion cerinorubellum* (Brauer, 1865)

Common name: Orange-tailed marsh dart

Description: In male the eyes are dark olivaceous in upper region and in lower region they are pale green. The thorax region is green in upper dims to blue colour on the sides and the lower region is yellow in colour. Legs are yellowish in colour which has small black spines. Wings are transparent and wing spot is amber coloured. The abdomen is brick red in colour while the middle segments are blue grey which has black bands. Female are identical to that of males but the abdominal segments are dull red in colour.

Size: The abdomen of the male is about 31to 33mm and the hind wing ranges from 20 to 21mm. In female the abdomen is about 31-35mm and the hind wing is from 20 to 21mm.

Habit and habitat: They are usually observed near rivers, ponds, lakes and in running waters. They are found as relaxed colonies and fly at medium heights in the vegetation. They are good predators of mosquitoes and are observed in the same marsh for long period of time.

Distribution: Found in various parts of Tiruvallur district.

17. Species: *Ceriagrion coromandelianum* (Fabricius, 1798)

Common name: Coromandel marsh dart

Description: In male the eyes are olivaceous in the upper region and in the lower region they are dim green in colour. The upper region of thorax is olive green in colour which merge to yellowish colour in sides and whitish in colour in its lower side. Legs are yellowish in colour which has spines in black colour. Wings are transparent and have a golden yellow spot in wing. Abdomen is bright yellow in colour. In female the thorax region is golden brown which gets dimmer on the sides. Abdomen is evenly olivaceous or golden brown colour at the back side.

Size: The abdomen of the male is about 28 to 30mm and the hind wing ranges from 18 to 20mm. In female the abdomen is about 29-32mm and the hind wing is 20mm.

Habit and habitat: They are found near rivers, ponds, lakes, temporary and permanent water bodies. They are weak fliers but males exhibit aggressive behavior in territory and are voracious feeders on midges and flies.

Distribution: Found in various parts of Tiruvallur district.

Genus: *Pseudagrion* (Selys, 1876)

18. Species: *Pseudagrion microcephalum* (Rambur, 1842)

Common name: Blue dart

Description: In male the eyes have a brown covering in the upper and a deep azure blue which gets dim to sky blue colour. The thorax region is azure blue in colour which has expanded black stripes. Legs are azure in colour. Wings are transparent and have a grey coloured wing spot. Abdomen is azure blue in colour of which the second segment has black colour marking on the upper part where the black markings are expanded from the segment third to seventh. Eighth segment has hard ring closes to the end and the ninth segment is devoid of markings. Tenth segment possesses a black mark which is saddle shape. In female the eyes are olive green colour in the upper region and dim blue in colour below. Thorax region is bluish green in colour and orange colour in the upper region and bluish colour on the sides. Wings are transparent and have a pale brown wing spot. In abdomen the second segment has a hard dumbbell shape above. In segments eight and nine, expanded black stripe are present in the upper region which has two spots. Tenth segment is devoid of markings.

Size: The abdomen of the male is 27mm and the hind wing is 17mm. In female the abdomen is 29mm and the hind wing is 20mm.

Habit and habitat: They are commonly observed in the vegetation near ponds, canals, rivers, lakes, temporary and permanent water bodies. They are mainly found in plains, and are one of the migratory species which migrates in high numbers.

Distribution: Found in various parts of Tiruvallur district.

Subfamily: Ischnurinae

Genus: *Ischnura* (Charpentier, 1840)

19. Species: *Ischnura aurora* (Brauer, 1865)

Common name: Golden dartlet

Description: In males the eyes are narrow which are black in colour in the form of a cap in the upper and the down region has a dark olive to olive green where in beneath it is a dim pale olive colour. A pair of black spots is seen on the back of the head. Thorax glowing black with a pair of dim green colour stripes while the sides of the thorax are dim green and with white colour below. Legs are dull greenish white in colour. Wings are transparent and the wing spot is red rose in colour in fore wings and dim grey in the hind wings. It has bright yellow abdomen and the second and seventh segment has an expansive black markings and the segments eight to ten are deep blue in colour presence of a black spot on the top of the tenth segment. Female is dull when compared to male with presence of a black stripe on the top side of the abdomen and the segments eight to ten are devoid of deep blue colour.

Size: The abdomen of the male is about 16-20mm and the hind wing ranges from 10 to 20mm. In female the abdomen is about 18 to 20mm and the hind wing from 14 to 15mm.

Habit and habitat: They are usually observed near ponds, rivers, lakes and temporary water bodies, and fly very close to the ground.

Distribution: Found in various parts of Tiruvallur district.

Superfamily: Calopterygoidea

Family: Calopterygidae

Subfamily: Calopteryginae

Genus: *Vestalis* (Selys, 1853)

20. Species: *Vestalis apicalis* (Selys, 1873)

Common name: Black-tipped forest glory

Description: In male the eyes are black with brown in the upper region and yellow with white in the bottom region. The upper and sides of the thorax are metallic green in colour and the underneath region is yellowish white in colour. The legs are black with brown. Wings are transparent and has a shining blue colour while the edge of the wings are blackish brown in colour. Abdomen is shimmering blue or green in colour which is black beneath. Female is identical to male but are dull when compared to markings on the wings which are pale, Abdomen is not shimmering as in male.

Size: The abdomen of the male is about 49 to 55mm and the hind wing ranges from 36 to 39mm. In female the abdomen is about 46 to 50mm and the hind wing from 38 to 40mm.

Habit and habitat: They are usually found near streams, hills, rivers and also found in the shaded areas of the bushes around the water habitat and are more common in forest areas.

Distribution: Found in various parts of Tiruvallur district.

21. **Species:** *Vestalis gracilis* (Rambur, 1842)

    Common name: Clear winged forest glory

    Description: In male the eyes are dark brown in the upper region and greenish yellow in the lower region. Thorax is metallic green in the upper region and yellow beneath. Legs vary from pale brown to dark brown in colour. Wings are transparent which has a shining blue colour. Abdomen is blue or green in the upper region while the lower region is black in colour. Female is similar to that of male while the abdomen is dim shimmering green in colour.

    Size: The abdomen of the male is about 45 to 46mm and the hind wing ranges from 34 to 38mm. In female the abdomen is about 43 to 50mm and the hind wing from 36 to 39mm.

    Habit and habitat: They are usually observed near hill streams, rivers, ponds and lakes, and are generally observed near the aquatic habitat which has fine shade pathways.

    Distribution: Found in various parts of Tiruvallur district.

Family: Platycnemididae

Subfamily: Platycnemidinae

Genus: *Copera* (Kirby, 1890)

22. **Species:** *Copera marginipes* (Rambur, 1842)

    Common name: Yellow bush dart

    Description: In male the eyes are black in upper, greenish on the sides with a black equatorial band beneath. Thorax is bronze black and the sides possess yellow lines. In the sides, the stripes are greenish-yellow in colour and are narrow. Legs vary from bright orange to reddish in colour. Wings are transparent and the wing spots are brown with yellow. Abdomen has a black bronze colour in the upper region, with stripes which are pale along the sides and at the end of the segments they form a pale green white colour ring. In female the eyes are identical to male but have olivaceous brown covering. Thorax is brown, and the black stripes are same as that of male and are irregular. Legs are brownish and have transparent wings. Abdomen is brown in the upper region which has broad rings towards the end.

    Size: The abdomen of the male is about 28 to 31mm and the hind wing ranges from 16 to

18mm. In female the abdomen is about 29 to 30mm and the hind wing is 20mm

Habit and habitat: They are found near ponds, rivers, canals, steams and slow flowing waters. They fly near water side vegetation at low heights.

Distribution: Found in various parts of Tiruvallur district.

5.5. Seasonal variation in abundance of dragonflies (Anisoptera) in the entire study habitat

Seasonal abundance of dragonflies in the Poondi habitat from January 2015 to December 2015 in all the seasons varied from 16.33 to 0.57 values. In the Poondi habitat, *Diplacodes trivialis* showed high abundance (16.33 ±3.05) in winter, (8.00 ±1.00) in summer and (11.00 ±1.00) in monsoon. In post monsoon, *Crocothemis servilia* showed high abundance (15.00 ±1.00). *Anax guttatus* showed low abundance in all the four seasons with (2.33 ±1.52, 0.33 ±0.57, 1.33 ±0.57 and 4.33 ±0.57 respectively) (Table: 7). ANOVA for the population abundance of dragonflies for the various seasons in Poondi habitat showed significance difference for the species like *Crocothemis servilia, Diplacodes trivialis, Neurothemis tullia, Pantala flavescens, Rhyothemis variegata, Trithemis aurora, Tholymis tillarga, Tramea limbata, Orthetrum Sabina, Anax guttatus* and *lctinogomphus rapax.* While the species *Brachythemis contaminata, Orthetrum pruinosum and Orthetrum glaucum* showed non significance (Table: 8).

Seasonal abundance of dragonflies in the Poondi habitat from January 2016 to December 2016 in all the seasons varied from 20.00 to 0.57. In the Poondi habitat, *Diplacodes trivialis* showed high abundance (18.33 ±3.05) in winter, (9.66 ±1.52) in summer, (14.00 ±1.00) in monsoon and (20.00 ±4.58) in post monsoon. *Anax guttatus* showed low abundance in all the four seasons (4.00 ±2.00, 0.66 ±0.57, 2.33 ±0.57 and 6.00 ±2.64 respectively) (Table: 9). ANOVA for the population abundance of dragonflies for the various seasons in Poondi habitat showed significance difference for the species like *Crocothemis servilia, Diplacodes trivialis, Neurothemis tullia, Orthetrum pruinosum, Pantala flavescens, Rhyothemis variegata, Trithemis aurora, Tholymis tillarga, Orthetrum glaucum, Orthetrum sabina, Anax guttatus* and *lctinogomphus rapax* while *Brachythemis contaminata* and *Tramea limbata* showed non significance difference (Table: 10). Student's T test analysis of dragonflies abundance at Poondi station for four seasons showed significant difference (Table: 11).

Seasonal abundance of dragonflies in the Puzhal habitat from January 2015 to December 2015 in all the seasons varied from 16.66 to 0.33. In the Puzhal habitat *Diplacodes trivialis* showed high abundance (16.66 ±2.51) in winter, (7.66 ±1.52) in summer and (12.33

±1.52) in monsoon. In post monsoon, *Brachythemis contaminata* showed high abundance (15.33 ±2.08). *Anax guttatus* showed low abundance in all the four seasons (2.66 ±1.52, 0.33 ±0.57, 1.66 ±.57and 3.00 ±1.00 respectively) (Table: 12). ANOVA for the population abundance of dragonflies for the various seasons in Puzhal habitat showed significance difference for all the species recorded from the site (Table: 13).

Seasonal abundance of dragonflies in the Puzhal habitat from January 2016 to December 2016 in all the seasons varied from 16.00 to 0.57. In the Puzhal habitat *Diplacodes trivialis* showed high abundance (14.00 ±2.00) in winter, (6.66 ±1.52) in summer, (11.33 ±1.52) in monsoon and (16.00 ±2.64) in post monsoon. *Anax guttatus* showed low abundance in all four seasons (1.00 ±1.00, 0.33 ±0.57, 1.33 ±0.57and 3.33 ±1.52) respectively (Table: 14). ANOVA for the population abundance of dragonflies for the various seasons in Puzhal habitat showed significance difference for all the species recorded from the site (Table: 15). Student's T test analysis of dragonflies abundance at Puzhal station for four seasons in the two years shows no significant difference (Table: 16).

Seasonal abundance of dragonflies in the habitat around the Sholavaram lake from January 2015 to December 2015 in all the seasons varied from 17.00 to 0.57 values. In the Sholavaram habitat, *Diplacodes trivialis* showed high abundance (17.00 ±2.64) in winter, *Pantala flavescens* and *Diplacodes trivialis* with (7.66 ±1.52) in summer and *Diplacodes trivialis* (12.33 ±1.52) in monsoon. In post monsoon, *Brachythemis contaminata* showed high abundance (16.66 ±1.52). *Anax guttatus* showed low abundance in all the four seasons (2.66 ±1.15, 0.33 ±0.57, 1.66 ±0.57 and 3.00 ±1.00 respectively) (Table: 17). ANOVA for the population abundance of dragonflies for the various seasons in the habitat showed significance difference for all the species recorded from the site while only *Pantala flavescens* showed non significance difference (Table: 18).

Seasonal abundance of dragonflies in the habitat around the Sholavaram lake from January 2016 to December 2016 in all the seasons varied from 17.00 to 0.57. In the habitat, *Diplacodes trivialis* showed high abundance (14.33 ±3.51) in winter, (6.66 ±1.52) in summer, (11.00 ±2.00) in monsoon and (17.00 ±2.64) in post monsoon. *Anax guttatus* showed low abundance (1.00 ±1.00, 0.33 ±0.57, 1.00 ±1.00 and 3.00 ±1.00) in all the four seasons (Table: 19). ANOVA for the population abundance of dragonflies for the various seasons in the habitat showed significant difference for all the species recorded from the site (Table: 20). Student's T test analysis of dragonflies abundance at this station for four seasons in the two years showed no significant difference (Table: 21).

Seasonal abundance of dragonflies in the habitat around the Avadi lake from January 2015 to December 2015 in all the seasons varied from 12.66 to 0.57. In the habitat, *Diplacodes trivialis* showed high abundance (11.66 ±3.05) in winter, (6.66 ±1.52) in summer and *Orthetrum sabina* (9.33 ±1.52) in monsoon. In post monsoon, *Brachythemis contaminata* showed high abundance (12.66 ±1.54). *Ictinogomphus rapax* showed low abundance (1.66 ±1.52) in winter, *Tholymis tillarga* (0.33 ±0.57) in summer, (1.33 ±0.57) in monsoon and (4.66 ±0.57) in post monsoon respectively (Table: 22). ANOVA for the population abundance of dragonflies for the various seasons in the habitat showed significance difference for all the species recorded from the site (Table: 23).

Seasonal abundance of dragonflies in the habitat around the Avadi lake from January 2016 to December 2016 in all the seasons varied from 12.66 to 0.57. In the habitat, *Diplacodes trivialis* showed high abundance (10.00 ±2.64) in winter, (5.33 ±1.15) in summer, (8.00 ±1.00) in monsoon and (12.66 ±2.08) in post monsoon. *Ictinogomphus rapax* showed low abundance (1.00 ±1.00) in winter, *Tholymis tillarga* (0.33 ±0.57) in summer, (1.00 ±1.00) in monsoon and (3.00 ±1.00) in post monsoon respectively (Table: 24). ANOVA for the population abundance of dragonflies for the various seasons in the habitat showed significance difference for all the species recorded from the site (Table: 25). Student's T test analysis of dragonflies abundance at this station for four seasons in the two years showed no significance difference (Table: 26).

Seasonal abundance of dragonflies in the habitat around the Ambattur lake from January 2015 to December 2015 in all the seasons varied from 14.33 to 0.57. In the habitat, *Diplacodes trivialis* showed high abundance (14.33 ±3.51) in winter, (8.66 ±1.52) in summer and (11.66 ±0.57) in monsoon. In post monsoon, *Crocothemis servilia* showed high abundance (13.33 ±1.52). *Ictinogomphus rapax* showed low abundance (3.66 ±1.52) in winter, *Trithemis aurora* (1.00 ±1.00) in summer, (1.66 ±1.15) in monsoon, *Crocothemis servilia* and *Orthetrum sabina* (13.33 ±1.52) in post monsoon respectively (Table: 27). ANOVA for the population abundance of dragonflies for the various seasons in the habitat showed significance difference for all the species recorded from the site (Table: 28).

Seasonal abundance of dragonflies in the habitat around the Ambatur lake from

January 2016 to December 2016 in all the seasons varied from 14.66 to 0.57. In the habitat, *Diplacodes trivialis* showed high abundance (13.00 ±2.64) in winter, (7.66 ±1.52) in summer, (11.33 ±0.57) in monsoon and (14.66 ±2.08) in post monsoon. *Tholymis tillarga* showed low abundance (4.33 ±1.52) in winter, (1.33 ±1.52) in summer, (1.33 ±1.52) in monsoon and (7.33 ±1.54) in post monsoon (Table: 29). ANOVA for the population abundance of dragonflies for the various seasons in the habitat showed significant difference where P< 0.05 value for all the species recorded from the habitat (Table: 30). Student's T test analysis of dragonflies abundance at this station for four seasons in the two years showed no significant difference (Table: 31).

Seasonal abundance of dragonflies in the habitat around the Korattur lake from January 2015 to December 2015 in all the seasons varied from 13.66 to 0.57. In the habitat, *Orthetrum sabina* showed high abundance (11.00 ±2.64) in winter, (7.33 ±1.15) in summer, (10.00 ±2.64) in monsoon and (13.66 ±2.51) in post monsoon. *Ictinogomphus rapax* showed low abundance (1.00 ±1.00) in winter, *Tholymis tillarga* and *Ictinogomphus rapax* (0.33 ±0.57) in summer, *Tholymis tillarga* (1.33 ±1.52) in monsoon and (3.00 ±1.00) in post monsoon (Table: 32). ANOVA for the population abundance of dragonflies for the various seasons in the habitat showed significant difference for all the species except *Brachythemis contaminata* and *Tholymis tillarga* recorded from the site. *Brachythemis contaminata* and *Tholymis tillarga* showed non significance difference (Table: 33).

Seasonal abundance of dragonflies in the habitat around the Korattur lake from January 2016 to December 2016 in all the seasons varied from 13.66 to 0.33. In the habitat, *Diplacodes trivialis* and *Orthetrum sabina* showed high abundance (9.33 ±1.52) in winter, *Orthetrum sabina* (6.00 ±1.00) in summer, (8.66 ±1.15) in monsoon and (13.66 ±1.52) in post monsoon. *Trithemis aurora* showed low abundance (1.66 ±1.15) in winter, *Rhyothemis variegata*, *Trithemis aurora*, *Tholymis tillarga* and *Ictinogomphus rapax* (0.33 ±0.57) in summer, *Tholymis tillarga* (1.00 ±1.00) in monsoon and (2.33 ±0.57) in post monsoon (Table: 34). ANOVA for the population abundance of dragonflies for the various seasons in the habitat showed significant difference for all the species except *Tholymis tillarga* (Table: 35). Student's T test analysis of dragonflies abundance at this station for four seasons in the two years showed no significant difference (Table: 36).

Seasonal abundance of dragonflies in the habitat around Poonamallee from January 2015 to December 2015 in all the seasons varied from 13.66 to 0.57. In the habitat,

Diplacodes trivialis showed high abundance (13.66 ±3.05) in winter, (8.66 ±2.30) in summer, (9.66 ±1.52) in monsoon and (11.66 ±1.52) in post monsoon. *Orthetrum glaucum* and *Trithemis aurora* showed low abundance (3.33 ±1.52) in winter, *Tholymis tillarga* (0.33 ±0.57) in summer, (1.33 ±0.57) in monsoon and (4.33 ±1.15) in post monsoon (Table: 37). ANOVA for the population abundance of dragonflies for the various seasons in the habitat showed significant difference for the species like *Brachythemis contaminata, Crocothemis servilia, Neurothemis tullia, Orthetrum pruinosum, Rhyothemis variegata, Trithemis aurora, Tholymis tillarga, Tramea limbata,* and *Orthetrum sabina. Diplacodes trivialis, Orthetrum glaucum* and *Pantala flavescens* did not show any significant difference (Table: 38).

Seasonal abundance of dragonflies in the habitat around Poonamalee from January 2016 to December 2016 in all the seasons varied from 13.00 to 0.57. In the habitat, *Diplacodes trivialis* showed high abundance (11.66 ±3.05) in winter, (7.00 ±1.73) in summer, *Crocothemis servilia* (8.33 ±1.15) in monsoon and *Diplacodes trivialis* (13.00 ±2.64) in post monsoon. *Trithemis aurora* showed low abundance (1.66 ±1.15) in winter, *Trithemis aurora* and *Tholymis tillarga* (0.33 ±0.57) in summer and *Tholymis tillarga* (0.66 ±0.57) in monsoon and (3.66 ±1.15) in post monsoon (Table: 39). ANOVA for the population abundance of dragonflies for the various seasons in the habitat showed significance difference for all the species recorded from the site (Table: 40). Student's T test analysis of dragonflies abundance at this station for four seasons in the two years showed no significance difference (Table: 41).

The seasonal abundance of dragonflies from the study habitats for the year 2015 and 2016 were shown in the Figure: 5 and Figure: 6 respectively.

5.6. Seasonal variation in abundance of damselflies (Zygoptera) in the entire study habitat:

Seasonal abundance of damselflies in the habitat around Poondi lake from January 2015 to December 2015 in all the seasons varied from 10.00 to 0.57. In the habitat, *Ceriagrion cerinorubellum* showed high abundance (7.66 ±3.21) in winter, (3.66 ±1.52) in summer and *Pseudagrion microcephalum* (7.66 ±0.57) in monsoon. In post monsoon, *Copera marginipes* showed high abundance (10.00 ±1.00). *Vestalis apicalis* showed low abundance (4.00 ±2.00) in winter, *Agriocnemis pygmaea* (1.00 ±1.00) in summer, *Vestalis gracilis* (3.33 ±1.52) in monsoon and *Copera marginipes* (10.00 ±1.00) in post monsoon (Table: 42). ANOVA for the population abundance of damselflies for the various seasons in Poondi habitat showed significance difference for the species *Copera marginipes, Vestalis gracilis, Agriocnemis pygmaea, Ceriagrion coromandelianum* and *Pseudagrion microcephalum* while *Vestalis apicalis, Ceriagrion cerinorubellum* and *Ischnura aurora* did not show any significant difference (Table: 43).

Seasonal abundance of damselflies in the habitat around Poondi lake from January 2016 to December 2016 in all the seasons varied from 12.66 to 1.00. In the habitat, *Ceriagrion coromandelianum* showed high abundance (9.33 ±3.05) in winter, (4.33 ±1.52) in summer, (8.00 ±1.00) in monsoon and (12.66 ±1.52) in post monsoon. *Vestalis apicalis* showed low abundance (3.66 ±2.08) in winter, (0.33 ±0.57) in summer, (2.66 ±1.52) in monsoon and (6.00 ±1.00) in post monsoon (Table: 44). ANOVA for the population abundance of damselflies for the various seasons in Poondi habitat showed significance difference value for all the species recorded from the habitat (Table: 45). Student's T test analysis of damselflies abundance at this station for four seasons in the two years showed no significance difference (Table: 46).

Seasonal abundance of damselflies in the habitat around Puzhal lake from January 2015 to December 2015 in all the seasons varied from 10.33 to 0.57. In the habitat, *Ceriagrion coromandelianum* and *Pseudagrion microcephalum* showed high abundance (9.00 ±2.00) in winter, *Ceriagrion coromandelianum* (4.33 ±1.15) in summer, (7.66 ±0.57) in monsoon and (10.33 ±0.57) in post monsoon. *Vestalis apicalis* showed low abundance (4.00 ±2.00) in winter, *Vestalis gracilis* (0.33 ±0.57) in summer, *Vestalis gracilis* and *Ischnura aurora* (4.33 ±0.57) in monsoon and *Ceriagrion coromandelianum* (10.33 ±0.57) in post monsoon (Table: 47). ANOVA for the population abundance of damselflies for the various seasons in Puzhal habitat showed significant difference for the species, *Vestalis gracilis, Agriocnemis pygmaea, Ceriagrion cerinorubellum, Ceriagrion coromandelianum, Ischnura aurora* and *Pseudagrion microcephalum* while the species *Copera marginipes* and *Vestalis apicalis* showed no significant difference (Table: 48).

Seasonal abundance of damselflies in the habitat around Puzhal lake from January 2016 to December 2016 in all the seasons varied from 10.00 to 0.57. In the habitat, *Ceriagrion coromandelianum* and *Pseudagrion microcephalum* showed high abundance (8.00 ±2.00) in winter, *Ceriagrion coromandelianum* (3.00 ±1.00) in summer, *Copera marginipes* (6.66 ±1.52) in monsoon and (10.00 ±1.00) in post monsoon. *Vestalis apicalis* showed low abundance (3.00 ±2.00) in winter, *Vestalis gracilis* (0.33 ±0.57) in summer, *Ischnura aurora* (2.66 ±2.51) in monsoon and (7.00 ±2.00) in post monsoon (Table: 49). ANOVA for the population abundance of damselflies for the various seasons in Puzhal habitat showed significant difference for the species *Copera marginipes, Vestalis gracilis, Agriocnemis pygmaea, Ceriagrion cerinorubellum, Ceriagrion coromandelianum* and *Pseudagrion microcephalum* while *Vestalis apicalis* and *Ischnura aurora* showed no significant difference (Table: 50). Student's T test analysis of damselflies abundance at this station for four seasons in the two years showed no significant difference (Table: 51).

Seasonal abundance of damselflies in the habitat around Sholavaram lake from January 2015 to December 2015 in all the seasons varied from 9.00 to 0.57. In the habitat, *Pseudagrion microcephalum* showed high abundance (6.66 ±1.52) in winter, *Copera marginipes* and *Pseudagrion microcephalum* (2.33 ±0.57) in summer, *Ceriagrion coromandelianum* (5.33 ±1.52) in monsoon and *Copera marginipes* (9.00 ±2.64) in post monsoon. *Vestalis apicalis* showed low abundance (2.33 ±1.52) in winter, *Vestalis apicalis* and *Ischnura aurora* (0.33 ±0.57) in summer, *Ischnura aurora* (2.33 ±0.57) in monsoon and *Copera marginipes* (9.00 ±2.64) in post monsoon (Table: 52). ANOVA for the population abundance of damselflies for the various seasons in Sholavaram habitat showed significant difference for all the species recorded from the habitat (Table: 53).

Seasonal abundance of damselflies in the habitat around Sholavaram lake from January 2016 to December 2016 in all the seasons varied from 9.33 to 0.57. In the habitat, *Pseudagrion microcephalum* showed high abundance (8.00 ±2.00) in winter, *Copera marginipes, Agriocnemis pygmaea* and *Ceriagrion coromandelianum* (3.00 ±1.00) in summer, *Copera marginipes* and *Ceriagrion coromandelianum* (6.33 ±1.52) in monsoon and *Ceriagrion coromandelianum* (9.33 ±0.57) in post monsoon. *Vestalis apicalis* showed low abundance (3.00 ±2.00) in winter, *Vestalis apicalis* and *Vestalis gracilis* (0.33 ±0.57) in summer, *Ischnura aurora* (3.33 ±0.57) in monsoon and *Ceriagrion coromandelianum* (9.33 ±0.57) in post monsoon (Table: 54). ANOVA for the population abundance of damselflies for the various seasons in Sholavaram habitat showed significant difference for all the species recorded from the habitat (Table: 55). Student's T test analysis of damselflies abundance at this station for four seasons in the two years showed no significant difference (Table: 56).

Seasonal abundance of damselflies in the habitat around Avadi lake from January

2015 to December 2015 in all the seasons varied from 10.00 to 0.57. In the habitat, *Pseudagrion microcephalum* showed high abundance (8.00 ±2.00) in winter, *Ceriagrion cerinorubellum* (5.00 ±2.00) in summer, *Agriocnemis pygmaea* (7.00 ±2.00) in monsoon and *Copera marginipes* and *Ceriagrion cerinorubellum* (10.00 ±1.00) in post monsoon. *Vestalis apicalis* showed low abundance (4.33 ±1.52) in winter, *Vestalis gracilis* (1.00 ±1.00) in summer, *Vestalis apicalis* (5.00 ±1.00) in monsoon and (5.66 ±0.57) in post monsoon (Table: 57). ANOVA for the population abundance of damselflies for the various seasons in the Avadi habitat showed significant difference for the species *Vestalis gracilis, Agriocnemis pygmaea* and *Pseudagrion microcephalum* while *Copera marginipes, Vestalis apicalis, Ceriagrion cerinorubellum* and *Ceriagrion coromandelianum* showed no significant difference (Table: 58).

Seasonal abundance of damselflies in the habitat around Avadi lake from January 2016 to December 2016 in all the seasons varied from 9.66 to 0.57. In the habitat, *Ceriagrion cerinorubellum* showed high abundance (6.33 ±3.05) in winter, (3.33 ±2.08) in summer, and (6.33 ±0.57) in monsoon and *Pseudagrion microcephalum* (9.66 ±1.52) in post monsoon. *Vestalis apicalis* showed low abundance (2.66 ±1.52) in winter, *Vestalis gracilis* (0.33 ±0.57) in summer, *Vestalis apicalis* (3.33 ±0.57) in monsoon and (6.00 ±1.00) in post monsoon (Table: 59). ANOVA for the population abundance of damselflies for the various seasons in Avadi habitat showed significant difference for all the species recorded from the habitat (Table: 60). Student's T test analysis of damselflies abundance at this station for four seasons in the two years showed no significant difference (Table: 61).

Seasonal abundance of damselflies in the habitat around Ambattur lake from January 2015 to December 2015 in all the seasons varied from 17.00 to 0.57. In the habitat, *Vestalis gracilis* showed high abundance (14.33 ±3.51) in winter, (6.66 ±1.52) in summer, (11.00 ±2.00) in monsoon and (17.00 ±2.64) in post monsoon. *Pseudagrion microcephalum* showed low abundance (4.33 ±1.52) in winter, *Agriocnemis pygmaea* (2.00 ±1.00) in summer, (4.00 ± 1.00) in monsoon and (8.00 ± 2.00) in post monsoon (Table: 62). ANOVA for the population abundance of damselflies for the various seasons in the habitat showed significant difference for all the species recorded from the habitat (Table: 63).

Seasonal abundance of damselflies in the habitat around Ambattur lake from January 2016 to December 2016 in all the seasons varied from 11.00 to 0.57. In the habitat, *Pseudagrion microcephalum* showed high abundance (7.66 ±1.52) in winter, (4.00 ±1.00) in summer, and (7.00 ±1.00) in monsoon and *Copera marginipes* (11.00 ±2.00) in post monsoon. *Agriocnemis pygmaea* showed low abundance (4.33 ±1.52) in winter, *Vestalis gracilis* (0.66 ±1.15) in summer (3.66 ±1.52) in monsoon and *Copera marginipes* (11.00 ±2.00) in post monsoon (Table: 64). ANOVA for the population abundance of damselflies for the various seasons in the habitat showed significant difference for all the species recorded from the habitat (Table: 65). Student's T test analysis of damselflies abundance at this station for four seasons in the two years showed no significant difference (Table: 66).

Seasonal abundance of damselflies in the habitat around Korattur lake from January 2015 to December 2015 in all the seasons varied from 9.33 to 0.57. In the habitat, *Ceriagrion cerinorubellum* showed high abundance (9.33 ±2.08) in winter, (4.33 ±1.15) in summer, *Ceriagrion coromandelianum* (6.66 ±1.52) in monsoon and *Pseudagrion microcephalum* (9.00 ± 1.00) in post monsoon. *Vestalis apicalis* showed low abundance (3.33 ±1.52) in winter, *Vestalis gracilis* (0.33 ±0.57) in summer, *Vestalis apicalis* and *Vestalis gracilis* (4.33 ±1.52) in monsoon and *Pseudagrion microcephalum* (9.00 ±1.00) in post monsoon (Table: 67). ANOVA for the population abundance of damselflies for the various seasons in the habitat showed significant difference for all the species recorded from the habitat (Table: 68).

Seasonal abundance of damselflies in the habitat around Korattur lake from January 2016 to December 2016 in all the seasons varied from 9.33 to 0.57. In the habitat, *Ceriagrion cerinorubellum* showed high abundance (8.33 ±2.08) in winter, (3.00 ±1.00) in summer, *Ceriagrion coromandelianum* (5.66 ±1.52) in monsoon and *Copera marginipes* (9.33 ±1.52) in post monsoon . *Vestalis apicalis* showed low abundance (2.33 ±1.52) in winter, *Vestalis apicalis* and *Vestalis gracilis* (0.66 ±0.57) in summer and *Vestalis gracilis* (2.33 ±1.52) in monsoon and *Vestalis apicalis* (6.66 +1.52) in post monsoon (Table: 69). ANOVA for the population abundance of damselflies for the various seasons in the habitat showed significant difference for all the species recorded from the habitat (Table: 70). Student's T test analysis of damselflies abundance at this station for four seasons in the two years no significant difference (Table: 71).

Seasonal abundance of damselflies in the habitat around Poonamallee from January 2015 to December 2015 in all the seasons varied from 10.00 to 0.57. In the habitat, *Pseudagrion microcephalum* showed high abundance (8.00 ±2.00) in winter, *Ceriagrion cerinorubellum* (3.66 ±1.52) in summer, *Pseudagrion microcephalum* (7.66 ±0.57) in monsoon and *Copera marginipes* (10.00 ±1.00) in post monsoon. *Vestalis apicalis* showed low abundance (4.00 ±2.00) in winter, *Vestalis gracilis* (1.66 ±0.57) in summer and (3.33 ±1.52) in monsoon and *Vestalis apicalis* (5.66 ±1.52) in post monsoon (Table: 72). ANOVA for the population abundance of damselflies for the various seasons in the habitat showed significant difference for all the species recorded from the habitat (Table: 73).

Seasonal abundance of damselflies in the habitat around Poonamallee from January 2016 to December 2016 in all the seasons varied from 9.66 to 0.57. In the habitat, *Ceriagrion cerinorubellum* showed high abundance (6.00 ±2.64) in winter, *Copera marginipes* and *Pseudagrion microcephalum* (2.00 ±1.00) in summer, *Pseudagrion microcephalum* (4.66 ±0.57) in monsoon and *Ceriagrion cerinorubellum* (9.66 ±2.08) in post monsoon. *Vestalis apicalis* showed low abundance (2.66 ±2.51) in winter, (0.66 ±0.57) in summer and *Vestalis gracilis* (2.66 ±1.52) in monsoon and *Vestalis apicalis* (7.66 ±2.08) in post monsoon (Table: 74). ANOVA for the population abundance of damselflies for the various seasons in the habitat showed significance difference for all the species except *Copera marginipes* which showed no significant difference (Table: 75). Student's T test analysis of damselflies abundance at this station for four seasons in the two years showed no significant difference (Table: 76).

The seasonal abundance of damselflies from the study habitats for the year 2015 and 2016 were shown in the Figure: 7 and Figure: 8 respectively.

5.7. Regression analysis:

Regression analysis for various physical parameters and there relation with the abundance of the dragonflies for the year 2015 were analyzed (Table: 77) and for 2016 were also analyzed (Table: 78). The environmental factors such as Temperature and RH showed positive correlation while the rainfall and wind speed showed moderate positive correlation. There is no negative correlation was observed for the environmental factors in all study habitat. The various environmental factors and their influence in the diversity of dragonflies in the study habitats for both the years was analyzed (Figure: 9) to (Figure: 22). Among damselflies in the year 2015 temperature and RH showed positive correlation where the rainfall and wind speed showed moderate positive correlation. But in the year 2016 the Temperature showed positive correlation while the RH, rainfall and wind speed showed moderate correlation. Regression analysis for various physical parameters and there relation with the abundance of the damselflies was shown for the year 2015 (Table: 79) and for 2016 (Table: 80). The various environmental factors and their influence in the diversity of damselflies in the study habitats for both year 2015 and 2016 were analyzed (Figure: 23) to (Figure: 36). There is no negative correlation was observed among the environmental factors in the study habitats.

Conservation Strategies:

Odonates are often exposed to wide range of pollutants in their natural habitats, but the understanding of their impacts on odonates needs more attention. There should be proper measures to be adopted to control the pollution causing agents. Most wetlands and rivers are bordered by highly transformed agricultural land with intensified farming systems that rely heavily on pesticides and fertilizers affecting the diversity and composition of odonates assemblages. Shifting to organic farming will reduce harmful toxic run off and fertilizer induced eutrophication which will help in odonates conservation.

Conservation of freshwater protected areas is mainly focused on large vertebrates mammals, birds and plants, which to some extent are umbrella taxa for other organisms, including macroinvertebrates. Thus, the integration of the key lotic habitats into the list of protected areas is fundamental for the conservation of odonates, and it should involve the maintenance of the integrity of the riparian zones. Macroinvertebrates in general are interesting bioindicators, providing reliable information on environmental disturbance faster than other taxa because of their shorter life cycles. Integrating macroinvertebrates into biodiversity monitoring schemes could generate new streams of information that complement data on other taxa, which will be useful for the conservation of biodiversity.

Encouraging and developing the scientific research in Odonatology, the natural history, ecology and conservation of the species, habitats and ecosystems in which odonates coexist. Conducting programs and awareness to the governments, public and private sector organizations on the importance of odonata conservation and its impact on environment.

Enlighten the people about the importance of odoantes and its impact on environment by conducting awareness programs, training programs, workshops and conferences. Conducting systematic surveys and research in sensitive regions to study odonates and its biodiversity status. Establishing awards and scholarships to individuals and organizations in recognition of their contribution to odonata conservation and its research.

6.0. DISCUSSION

Biodiversity of a particular habitat includes distribution, abundance, composition, cooperation of various species and functional landscape in the ecosystem. All these components together form a healthier ecosystem. Diversity in a specific habitat depends on the various factors like temperature, humidity and available of food sources which play an important role in maintaining ecosystem (Hunter, 2002). In a natural environment, diversity plays a significant role by maintaining a healthy environment. Nowadays, in ecology the study on biodiversity has received main focus on most of the countries since they have adopted bio monitoring measures in the various habitats to know their status and the various factors affects them (Yoccoz *et al.,* 2001). The diversity in the specific habitat includes various assemblages of flora and fauna, whenever there is little variation in it which not only affects the specific habitat but also the entire ecosystem. Biodiversity has not received main focus even though they are important wealth of the country. Nowadays none have a correct understanding that how much species are present till date on the planet. Many have an opinion that global species varies from 3 to 100 million (Erwin, 1982; Stork, 1988; Hodkinson and Casson, 1991; May, 1992; Raven and Wilson, 1992). The study of species in habitat and their composition is very much necessary to estimate the overall biodiversity.

Hooper *et al.,* (2005) stated that in recent years the relationship between biodiversity and ecosystem functions is a topic generated intensive research. Odonata has been considered as an excellent, easy to use indicator group to understand the status of biodiversity in freshwater ecosystem (Clausnitzer, 2003; Suhling *et al.,* 2006). In the present study, the diversity of odonates recorded in seven habitats of Tiruvallur district in Tamil Nadu (Poondi, Puzhal, Sholvaram, Avadi, Ambattur, Korattur and Poonamallee) was carried from January 2015 to December 2016. In the entire study period, a total 14 species of dragonflies under three families and eight species of damselflies under three families comprising a total 22 species of odonates were identified in all the study sites. In dragonflies, family Libellulidae was the most dominant with five species followed by one species from family Aeshnidae and family Gomphidae. In case of damselflies family Coenagrionidae showed the highest dominance with five species followed by family Calopterygidae and family Platycnemididae with two and one species respectively.

Abundance of the odonates recorded during the study showed variance among the study sites. Among the study sites, the Poondi habitat showed rich natural vegetation and the habitat did not support most of the pollution causing activities. High diversity of Odonates was observed in this habitat. Likewise the habitat around the water bodies which supports good natural vegetation and which has less chances for the pollution causing activities showed high diversity of odonates. The habitat around the Puzhal lake, Sholavaram lake also showed rich diversity of odonates. The habitat around the Avadi, Ambatur and Koratur lake showed a good diversity while the Poonamallee habitat showed a moderate diversity. The presence of odonates depends on the good aquatic habitat and availability of plants near the water bodies. The influence of green vegetation plays an important role in the diversity of odonates in many habitats (Clark and Samways 1996; Stewart and Samways 1998; Schindler *et al.*, 2003; Hofmann and Mason, 2005). The availability of reeds gives necessary habitat for the nymph and adults which increases the diversity of odonates (Samways and Steytler, 1996). For oviposition, the reeds provide necessary sites for the odonates and they are used by the nymph in their final instar and while coming out of water they climb on it for drying (Thompson *et al.*, 2003; Rouquette and Thompson, 2007) and also for shelter and hiding. Since the habitat around the Avadi, Ambattur and Koratur lake has been highly explored to high anthropogenic activities, these have led to the destruction of the habitat. Koratur lake was highly affected by various human activities and also suffers from eutrophication. Industrial waste and sewage inflow also affects the lake habitat which leads to the destruction in diversity of odonates.

Subramanian (2005) studied that odonate species are highly habitat specific. Habitat quality plays a vital role in distribution of odonates in an aquatic habitat and healthy aquatic habitat shows a rich diversity of odonates while a destructed aquatic habitat shows a less diversity of odonates. Poonamallee habitat exhibited a moderate diversity of odonates since it has been rapidly developing leading to destruction of natural habitat and water bodies. Dolny *et al.*, (2011) studied that the Bornean rainforest where the anthropogenic activities in a habitat caused destruction of the species in that habitat. Human disturbances widely affect diversity of odonates and most of the areas in the habitat had high human disturbances which showed less diversity of odonates, where the habitat which was free from pollution and the habitat which had less human disturbances showed a high diversity of odonates.

The species diversity in a specific habitat is appearing to be one of the most important resources of an ecosystem. In an ecological system, it is necessary to figure out the diversity of organisms and various factors affecting them. A healthy habitat plays a vital role in the occurrence of certain species. Since odonates are habitat specific and are mainly found near aquatic habitats.

Seasonality plays a vital role in the diversity of insects and depending upon the seasons, the insect population varies which were observed in the current study. During the post monsoon, the population of odonates increased and they showed a high abundance till the winter seasons. During warm temperature particularly in the peak of the summer seasons the population of odonates was declined. Fluctuations in the rainfall had a huge impact on the availability of the tropical insects (Hill *et al.*, 2003). Humidity appears to be in relation with the odonata diversity. The humidity in the ecosystem provides a favourable condition for the density of odonates. Monsoon had high impact on the distribution of odonates. Maximum diversity of odonates was observed in the post monsoon. Thus the Monsoon provides a favourable platform for the growth of various kinds of plants and also for insects. The availability of water which provides humidity and also the prey for odonates leads to increase in the diversity of odonates.

Temperature had an adverse impact on the distribution and diversity of odonates. In the current study, temperature certainly interacted with the diversity and species richness of odonates. Moderate temperature provides favorable environment for the odonates. During the summer, there was low humidity and high temperature which had a high impact on the diversity of odonates. In summer, due to tremendous heat scarcity of water occurs which certainly have adverse effect on the growth of insects on which odonates feed. Humidity, rainfall and temperature play a vital role in determining the diversity of odonates. In the entire study area, high diversity of odonates were observed when there was rich natural vegetation and pleasant climatic factors.

Dragonflies prefer sunny habitat to maintain their body temperature and they are observed in high diversity in the sunny habitat while damselflies showed high diversities in the area which possess shade cover. Dragonflies regulate their body temperature ectothermically by control of solar input and thermoregulation is achieved by behavior and physiological responses (Corbet, 1999). Further, most dragonflies prefer sunny areas with low woody vegetation (Clark and Samways, 1996). Most of the dragonflies are diurnal but a few actively hunt during twilight hours. They feed in flight, using the legs to capture the prey and transfer it to the jaws. The legs are highly specialized for this purpose, particularly with regard to their position, relative length, articulation and complement of spines. The vision of dragonflies is well developed and the eyes are very large in proportion to the head. Vision dominates their behavior, including searching for prey and looking for mates.

In all the study sites between the two suborders, Anisoptera was found to be dominant than the Zygopetra. The maximum diversity and abundance observed in Anisoptera where compared to Zygopetra. Their higher distribution may be due to their huge dispersion capacity (Kadoya *et al.*, 2004) and their tolerance to large habitats (Suhling *et al.*, 2005). Most of the diversity was observed in the areas which had good open sunny areas. Light source plays a vital role in selection of habitat. In the present study, species of family Libellulidae (Anisoptera) was dominant in all the study sites. They are widely distributed in all the study sites with 12 species. Since these species are tolerance to a wide range of habitats, they are able to occupy all the habitats in large number. Since they are adapted to a wide geographical area over the lentic habitat and their small period of life cycle makes them ideal over all the areas (Subramanian *et al.*, 2008). The lentic habitat shows greater diversity of odonates. Odonates depend on the water bodies for their reproduction. After mating, female lays their eggs on water or submerged plants and the larval stages occurs in water (Hornung and Rice, 2003), but adult odonates exhibit aerial mode of life. Odonates mainly depend on the waters for their food and reproduction and are found accumulated near water habitats (Oppel, 2005). In the entire study, it was observed that a large number of species were found near marsh land and the habitats around the lake. Larva of the odonates were found widely in the lentic waters, while the adult odonates were seen near the water bodies.

In the study sites, seven species of damselflies (Zygoptera) were observed. Among them, family Coenagrionidae showed more dominance with four species in all the study sites. It is one of the common families, due to their short period of life cycle and tolerance to wide range of habitats they are able to occupy all the habitats in large number (Gentry *et al.*, 1975). Damselflies are delicate insects and they are weak flyers. Zygoptera species reflected greater ecological diversity and they were recorded at sites with higher percentage shade. Shade and rich aquatic habitat provides favourable condition for zygopterans. The have large eyes that are distinctly separated occupying both sides of the head. The adults will catch their prey with the help of their legs and they contain chewing kind of mouth parts but in the larval stage they are voracious feeders and it has forbidden mouth parts that precisely engulf the prey (Tiple *et al.*, 2012).

High assemblages of damselfly are related with lengthy, narrow and diverse littoral and lacustrine zones. The aquatic plants around the habitat are necessary for the various activities like foraging, shelter and oviposition. Damselflies lay their eggs on the aquatic plants which make them structure specific and species specific (Corbet, 1999). Damselflies are found in abundance in the natural vegetation which provides a shade for them. Damselfly species are found in large amount in the areas of dense forests (O" Neill and Paulson, 2001) and they are also seen in shade part of streams (Brooks and Jakson, 2001). Due to their small body size they may prefer shaded areas, and are considered as thermal conformers (May, 1976), showing maximum conductance with their temperature differing in environmental temperature. In the four seasons, post monsoon provides favorable condition for odonates, but summer provides unfavorable condition for their diversity. During the present study, maximum species richness and diversity of odonates were recorded during the post monsoon. In the post monsoon, due to availability of rich source of plant vegetation and high water sources, they provide a favorable condition for the odonates. In summer, unavailability of water affects the odonates survival since they rely on water bodies for their life cycle.

Odonates are predators both in their larval and adult stages they play a vital role in the wetland ecosystem. Adult odonates feed on mosquitoes, black flies and other blood-sucking flies and serve as a significant bio-control agent. Most species of odonates occupy the agro ecosystems and perform an important role in the control of pests. Their larvae act as a natural bio-control agent for the mosquito larvae and thus control several epidemic diseases like malaria, dengue and filarial (Mitra, 2002). Odonates are the first arthropods considered as biological control agents in controlling mosquitoes due to their colonization, management and handling made them difficult in adapting them (Legner, 1995). Dragonflies are used to control mosquitoes and they are used in the pest management (Sathe and Shinde 2008).

In an environment, diversity of species in a specific habitat is the most important property. It is necessary in an environment to know the diversity of the organisms and factors that regulate it (Apodaca and Chapman, 2004). To establish an effective ecosystem, the biodiversity richness plays a vital role in it. It is important to provide rich biodiversity other than providing other benefits (Tilman, 1999; Purvis and Hector, 2000). In recent days, the biodiversity has been facing a lot of problems due to increase in population, overuse of the natural resources, increase in the pollution and destruction of the natural habitat which leads to destruction of the natural ecosystem. Nearly, all the species and organisms occupying the same tropical level in the ecosystem play a vital role in maintaining the balanced ecosystem. Thus, the biodiversity has been classified as within the species, among the species and among the ecosystem (Heywood and Watson, 1995).

To trace the health of any habitat the diversity measurement of that habitat is considered to be important (Magurran, 1988). Any change in the habitat which affects the habitat, it shows the reduction of species in that habitat. In the study sites, due to rapid development, a change in the habitat clearly affected the diversity of odonates. The year 2015 showed a higher diversity than 2016. When the habitat is reduced qualitatively and quantitatively it showed a decline in the diversity of the species (Jenkins, 2003).

Diversity index is studied by using the Shannon-Weiner Index which has been derived by Shannon and Weiner and known as Shannon Diversity Index (Krebs, 1985). Magurran (1988) stated that the individuals collected from the high population of the habitat and that all species should be representing in it. May (1976) stated Shannon Weiner index value as normally when they occur from 1.5-3.5 diversity of species for the ecological data and hardly exceeds 4.0 while the Simpson Index ranges between 1-0 and where 1 shows diversity and 0 shows no diversity.

Ecosystem which have a wide range of diversity can tolerate the environmental stress to a greater extent while the habitat which suffers from loss of species has only little capacity to withstand or retrieve from the deterioration. It is in the starting stage to know in depth the measures that help to maintain the global biodiversity (Storch *et al.*, 2007). Ecological indicators are the community or specific taxa which help to know the status of the biotic or abiotic components of a habitat. Odonates adapt to aquatic habitats and they are used as good habitat indicators because of their habitat necessity (Corbet, 1999).

All group of water body has a characteristic species and association of odonates is generally found near the aquatic habitats. Odonates adapt themselves to aquatic habitat in the larval stage and the adult stage prefer terrestrial habitat. They are sensitive to certain habitat and are used to analyze any environmental damages in aquatic habitats. (Clark and Samways 1996;

Sahlén and Ekestubbe 2001; Clausnitzer 2003; Foote and Hornung 2005; Osborn 2005). They respond very well whenever there is an alteration in the aquatic habitat and also to any change in the water body (Balzan, 2012). Odonates have been extensively proposed as indicators of environmental quality in aquatic habitat. After copulation, these insects lay their eggs in or near the freshwater. So their huge abundance in a habitat is a good indicator of the quality of freshwater (Corbet, 1999). Odonata (dragonflies and damselflies) are used as bioindicators for wetland quality in Europe, Japan, the USA, Australia (Clausnitzer and Jodicke, 2004) and in South Africa (Stewart and Samways, 1998). Odonates are used as indicators they provide several advantages since they are widely spread and it have been one of the mostly studied groups. As habitat quality changes, odonates also exhibit changes in their diversity and distribution due to their sensitivity to structural habitat quality (Clark and Samways, 1996).

Dragonflies which are present in the uninterrupted habitats with rich riparian vegetation are specialists with narrow distribution, but the species which are observed at the industrial land or urban areas in the disturbed riparian vegetation were in general areas with large habitat distribution. The study also showed that dragonflies were sensitive not only to the quality of the wetland but also to the large habitat alteration, specifically in the riparian area. Due to increase human activities, the environment is adversely affected due to accumulation of various chemicals and toxins. Hence bioindicators are sensitive tools used for prediction and recognition of environmental stresses. Whenever environment is affected by pollutants the bioindicator indicates the unhealthy situation either by showing certain changes in their distribution or completely disappearing from the environment. The most important reasons for using bioindicators are the direct determination of biological effects and the effects of multiple pollutants in organisms. Insects like dragonflies and damselflies are used as excellent environmental markers to indicate the healthy status of water bodies by indicating their absence near highly polluted water bodies.

7.0. SUMMARY

Biodiversity includes genes in a population, various species in a community which maintains the balance in an ecosystem. Hence, it is necessary to document and enlist record of biodiversity. The invertebrates occupy the major part of the animal biomass on the planet earth, but are not given enough attention in the biodiversity assessment methods. Recently odonates are globally used for analyzing the status of the biodiversity as they indicators of fresh water habitat, and therefore any undesirable change in the habitat can be identified. They habitat near clean water ecosystem and high abundance of odonates are seen in an habitat which supports healthy biodiversity.

The diversity of dragonflies and damselflies were observed in the seven habitats of Tiruvallur district, Tamil Nadu, India from January 2015 to December 2016. The diversity and seasonal variation of odonates around the seven habitats *viz.*, Poondi lake, Puzal lake, Sholavaram lake, Avadi lake, Ambattur lake, Korattur lake and Poonamallee were studied. Diversity of odonates were studied by using the line transects method where the odonates were collected by sweep net technique. Regular observations were made on weekly basis in the study habitats. During the study, only matured odonates were collected, photographed and identified with the help of identification keys. In the present study, a total of 22 species which comprises of 14 species of dragonflies from three families and eight species of damselflies from three families were recorded. Within the three families in Anisoptera, family Libellulidae showed the highest dominance with ten genera and twelve species while in Zygoptera, the family Coenagrionidae was dominant with four genera and five species.

Among the seven habitats, the habitat around Poondi, Puzhal and Sholavaram lake supported a rich diversity of odonates with all the 22 species recorded from the habitat while other habitats showed absence of some species. In Anisoptera, *Diplacodes trivialis* showed the highest dominance, while the species *Anax guttatus and Ictinogomphus rapax* showed less abundance in most of the habitats. In Zygoptera, *Ceriagrion coromandelianum* and *Ceriagrion cerinorubellum* showed maximum dominance. The diversity of odonates was also analyzed using Simpson, Shannon-Wiener and Margalef indices. The seasonal variations of the odonates in the study habitat during the various season *viz.*, post monsoon, monsoon, winter and summer were studied and analyzed. In the post monsoon, high range of odonata diversity was observed. High assemblage of odonates were seen where there was a rich source of aquatic plants, but when there is undesirable change in the aquatic habitat particularly in the summer season makes the survival of the odonates tough. Hence low diversity of odonates was seen in the summer season.

Abiotic factors had a great impact on the diversity and abundance of odonates. The various environmental factors, *viz.*, temperature, humidity, rainfall will directly affect the diversity and distribution of insect populations. Distribution of odonates highly depends on temperature. Odonates are generally active in the midday, since they are highly depend on sunlight to maintain their body temperature. Dragonflies were observed highly in the sunny habitat while dragonflies preferred shade cover areas. The diversity of odonates and their habitats were affected also due to various anthropogenic activities, *viz.*, improper use of land, large scale habitat destruction, loss and damage to their breeding habitats by exhausting of the swamps, pollution of water bodies, sand mining and eutrophication. In agro-ecosystems, odonates are important biocontrol agents helping in controlling insect pest population, besides their rich diversity and habitat specificity which make them ideal tools for assessing freshwater ecosystem health. Dragonflies and damselflies have shown to be useful for nature management and conservation. Therefore, public awareness is necessary to conserve the suitable habitats of this economically important invertebrate organisms.

www.ingramcontent.com/pod-product-compliance
Lightning Source LLC
LaVergne TN
LVHW051310200726
843510LV00010B/1361